MARY SOMERVILLE AND THE CULTIVATION OF
SCIENCE, 1815 – 1840

ARCHIVES INTERNATIONALES D'HISTOIRE DES IDEES
INTERNATIONAL ARCHIVES OF THE HISTORY OF IDEAS

102

MARY SOMERVILLE AND THE CULTIVATION OF SCIENCE, 1815 – 1840

ELIZABETH CHAMBERS PATTERSON

MARY SOMERVILLE AND THE CULTIVATION OF SCIENCE, 1815 – 1840

by

ELIZABETH CHAMBERS PATTERSON

1983 **MARTINUS NIJHOFF PUBLISHERS**
a member of the KLUWER ACADEMIC PUBLISHERS GROUP
BOSTON / THE HAGUE / DORDRECHT / LANCASTER

Distributors

for the United States and Canada: Kluwer Boston, Inc., 190 Old Derby Street, Hingham, MA 02043, USA
for all other countries: Kluwer Academic Publishers Group, Distribution Center, P.O.Box 322, 3300 AH Dordrecht, The Netherlands

Library of Congress Cataloging in Publication Data

```
Patterson, Elizabeth Chambers.
   Mary Somerville and the cultivation of science,
1815-1840.

   (Archives internationales d'histoire des idées =
International archives of the history of ideas ; 102)
   Bibliography: p.
   Includes index.
   1. Somerville, Mary, 1780-1872.  2. Women scientists
--Great Britain--Biography.  3. Science--Great Britain--
History.  I. Title.  II. Series: Archives internationales
d'histoire des idées ; 102.
Q143.S72P37  1983     500.2'092'4 [B]     83-2382
ISBN 90-247-2823-1
```

Q
143
.S72
P37
1983

ISBN 90-247-2823-1 (this volume)
ISBN 90-247-2433-3 (series)

Copyright

PRINTED IN THE NETHERLANDS

CONTENTS

ACKNOWLEDGEMENTS

Nearly 30 years ago the late Sir Brian Fairfax-Lucy, Bt. and his wife, the well-known writer and friend of scholarship, the Hon. Lady Alice Fairfax-Lucy, deposited at Somerville College, Oxford, a large collection of letters, papers and memorabilia that had come to him as the Fairfax heir. All had belonged to that remarkable nineteenth-century scientific lady, Mrs. Mary Fairfax Greig Somerville (1780–1872), after whom Somerville College is called. In time this material was transferred to the Bodleian Library, where in 1967 the author began, as a first step in its exhaustive study, the sorting, arranging and cataloguing of the more than 5000 items which now make up the Somerville Collection. This Collection and other pertinent pieces are the basis of this book, the first detailed scrutiny yet made of any part of Mrs. Somerville's life and accomplishments.

The writer is deeply indebted to the Fairfax-Lucys for permission to use the Somerville Collection and for their many kindnesses to her. She is grateful also to the Fellows of Somerville College for their interest and encouragement and for their warm hospitality during her stays in Oxford. Particular thanks are due the former Librarian of Somerville, Norma Lewis, now Lady Dalrymple-Champneys, who first brought the Collection to the author's attention and has since been a steadfast guide in its exploration, and to two recent Principals of the College, Dame Janet Vaughan, F.R.S. and Mrs. Barbara D. Craig, who also have given unstintingly of their time, knowledge and sympathy. The American Philosophical Society through two grants supported the cataloguing and microfilming of the papers, an invaluable assistance that is gratefully acknowledged.

Many great libraries, their librarians, archivists and staffs have aided in the searches on which this book is based. Special thanks are due the Bodleian Library, the National Library of Scotland, the British Library, the University of St. Andrews Library, Trinity College (Cambridge) Library, the Yale University Library, and the libraries of the Royal Society and of the Royal Institution.

Mrs. Christina Colvin of Oxford called attention to references to Mary Somerville in many unpublished letters of the Edgeworth family and permitted this investigator to examine them. Mr. John Robert Murray supplied from the Murray Archives copies of many Somerville letters and numerous details relating to the publication of Mrs. Somerville's books by John Murray I and John Murray II.

Prof. John D. North of the University of Groningen has been magnanimously generous with help at every stage of this project. Dr. A. C. Crombie of Trinity College, Oxford, has been a valued adviser and friend throughout the undertaking.

But my principal debt is to my husband, Prof. Andrew Patterson of Yale University, without whose wise encouragement and patient assistance in all matters large and small there should have been no book.

New Haven, Connecticut
2 March 1982

INTRODUCTION

Among the myriad of changes that took place in Great Britain in the first half of the nineteenth century, many of particular significance to the historian of science and to the social historian are discernible in that small segment of British society drawn together by a shared interest in natural phenomena and with sufficient leisure or opportunity to investigate and ponder them. This group, which never numbered more than a mere handful in comparison to the whole population, may rightly be characterized as 'scientific'. They and their successors came to occupy an increasingly important place in the intellectual, educational, and developing economic life of the nation. Well before the arrival of mid-century, natural philosophers and inventors were generally hailed as a source of national pride and of national prestige. Scientific society is a feature of nineteenth-century British life, the best being found in London, in the universities, in Edinburgh and Glasgow, and in a few scattered provincial centres.

During the years 1815 through 1840 — the span of this study — the most evident changes in British science were the emergence of a multitude of specialities within the broad field of natural philosophy and the formation of their attendant specialist learned societies; the emergence of a kind of primitive professionalism among scientific practitioners; and the emergence of wider and more substantial recognition and support of science by the government and by the nation, as the utility and promise of this branch of learning became manifest. At the beginning of this period, for example, only the Royal Society of London — embracing all the sciences of the day — and three specialist societies — the Linnean, the Horticultural, and the Geological — were firmly established. By the end of the period, these four had been joined by ten new scientific organizations, three of which — the Astronomical, the Botanic, and the Geographical — were also designated 'Royal' and another of which — the British Association for the Advancement of Science — subsumed every recognized scientific speciality.

In 1815 science was considered by the public and by its practitioners an avocation. Natural philosophers thought of themselves as gentlemen amateurs and regarded their organized societies as gentlemen's clubs. By 1840, however, a man could reasonably give all his working hours to science and honourably gain a livelihood thereby. Furthermore, proficients in various branches of science were by then slowly beginning to formulate suitable courses of training for novices and to set down acceptable standards of performance and other requirements for admission to their fields. Indiscriminate election of fellows to scientific societies was under heavy attack from within as demands grew to limit membership to gentlemen actively pursuing science and to regard election as a token of competence and esteem. Only the British Association could maintain in the coming years, as it had from its founding, a policy of admitting any man who wished to join.

In 1815 official recognition of scientific endeavours and of those engaged in them was scant and highly personal. Humphry Davy's knighthood in 1812 was the first conferred for scientific work since Isaac Newton's in 1705. But a growing insistence, chiefly by the scientific community, on signs of Royal approbation and on greater financial assistance in carrying out researches and establishing needed facilities resulted, from the 1830s, in readier bestowal of civil pensions and titles on scientific personages and in the funding of several large scientific undertakings. In an industrializing Britain, science and scientists found an increasingly important role. All these trends would intensify as time passed.

In the metropolitan centres of early nineteenth-century Britain science was a fashionable pastime. Crowds of stylishly dressed men and women flocked to lectures at the Royal Institution in London and elsewhere. Every smart drawing-room boasted a mineral cabinet. Scientific toys and small pieces of apparatus were commonplace on sitting-room tables and in ladies' boudoirs. A passion — be it ever so superficial — for science was modish in well-to-do circles. It was a deep and sincere interest, however, rather than mere fashion that brought a remarkable Scotswoman, Mrs. Mary Fairfax Greig Somerville, to the study of science and mathematics and guided her through her long and distinguished career as a scientific lady. Conditions and attitudes prevailing in the early nineteenth century made her career possible. A growing appreciation of science, an energetic nationalism, and her own considerable talents sustained it.

The period 1815 through 1840 marks Mrs. Somerville's life in London science, the years between her husband's initial appointment to a London post and the family's relinquishment — because of his ill health — of a permanent London residence. During this particular quarter century she was at the heart of the London scientific community, actively engaged in study, experimentation, and writing both there and in Paris. She was on familiar terms with the leading scientific men of her own country and of Europe, all of whom regarded her as a cherished colleague. With their encouragement she was able, through her own gifts and knowledge, to fashion a clear, concise and authoritative account of current science at a time when such a statement was lacking and much needed. Her *On the Connexion of the Physical Sciences* (1834) presented a comprehensive picture of the newest researches and ideas in the physical sciences, calling to the attention both of specialists and of general readers much recent scientific work hitherto little noticed but subsequently important.

Further, by her very choice of material for the book, Mrs. Somerville helped to define the then unsettled term 'physical sciences'. Throughout its many editions this volume retained its topical authority; no more reliable and convenient descriptions of the state of received physical science at the dates of its publication can be found. Earlier, through her mastery of French mathematics and her skill at exposition, she made an important contribution to the modernization of English mathematics through her *Mechanism of the Heavens* (1831). Often carelessly and foolishly dismissed by historians in the twentieth century as vapid popularizations of abstruse subjects by a mere woman, these seminal books were in their own times gratefully and enthusiastically received by scientific specialists, by many informed readers, and by a host of others whom they brought to science for the first time.

Largely self-educated until her arrival in London early in 1816, Mary Somerville served there, through good fortune buttressed by great talent and zeal, an informal apprenticeship under several of the foremost philosophers of the day. At that date no formal course of training had yet been designed or prescribed, and scientific men — safe from economic or professional threat from scientific women — were cordially welcoming to serious students be they male or female.

Nowhere could Mrs. Somerville have found better masters and her tutelege, similar in most respects to that given superior males, typifies the instruction then considered necessary and appropriate for a practising scientist. Her own experience, especially when contrasted with the difficulties experienced a few years earlier by an even more gifted aspirant to science Michael Faraday —young and of the labouring class — demonstrates the advantages derived from gentle birth, a 'proper' introduction into scientific circles, and previous and continuing ties to the cultivated and well-born. The readiness and ease with which Mrs. Somerville and her unusual interest in science were accepted by the London scientific community and the up-to-the-minute training she received from men who were at the very time creating the science of the day attests not only to the openness of scientific society to the talented and well-connected but also to the absence of any anti-female bias among scientists. That her apprenticeship and career flowered in London illustrates the importance of being at the centre of discovery. Mary Somerville was at the right place at the right time and had the ability and will to make good use of her situation.

The English Scientific Establishment in these decades was small and made up overwhelmingly of Fellows of the Royal Society. The importance of this association is emphasized in the text by the inclusion of the year of election along with the dates of birth and death of the chief persons in this study. Yet this Establishment, far from being monolithic, was beset with internal strains and differences arising principally from a new and gradually developing professionalism in science. Mary Somerville, liberal in all her views, was firmly attached to the reform elements in science, in society, and in politics. Nevertheless, through her sweetness of character and her good sense, she always maintained friendly and civil relations with conservative, even reactionary, circles. A strong, rational and compassionate woman, she was never a violent activist on behalf of change. Rather she preferred to work within the system, convinced that reason and pressing need for change would ultimately bring about transformations without sacrifice of valued traditions. Through her words and her example, she staunchly supported those causes she thought just, whatever popular opinion maintained.

Although on occasion Mrs. Somerville was criticized by unsympathetic persons — usually ill-informed or uncultured — for her 'unwomanly' pursuit of science, she also benefited greatly from the new respect accorded science and scientists in those years. At a time when national honours were largely reserved for the high-born and their connections, for politicians, or for military and naval figures and were rarely given intellectuals, she was handsomely recognized for her achievements by her country, her fellow countrymen and women, and by scientific colleagues. Before the middle of the century, she was accepted at home, abroad and overseas as the premier scientific lady of the ages. Before its end, a nation treasuring the image of itself as civilized, advanced, large-minded and innately superior had exalted her almost to the point of minor deification.

Adding greatly to her lustre in the eyes of her contemporaries was her exemplary fulfilment of her role as wife and mother. Displaying no tinge of blue, she moved easily and graciously in cultivated society, preserving all the traditional female traits and graces while adding to them extraordinary mental achievements. Moreover, she enjoyed the pleasures and gaiety of metropolitan life — music, art, the opera, the theatre, entertainments, excursions, and, above all, association with friends of like tastes. Sympathetic encouragement and useful assistance in her work from a husband who himself relished scientific society and gloried in her accomplishments played no small part in her career.

The present study considers only the scientific aspects of Mary Somerville's life from 1815 through 1840. Other factors, such as the Somervilles' financial circumstances, family crises, political conditions, international concerns and the like, necessarily enter but only insofar as they affect Mrs. Somerville's intellectual life and her work. Simultaneously with her pursuit of science, she carried on an active and interesting daily existence in the best literary and political society of London and, on occasion, abroad. An examination of the Somerville Collection and of manuscripts and printed accounts of many of her contemporaries reveals that during her long lifetime, Mary Somerville met and was frequently on familiar terms with many celebrated figures in her own country, France, Italy and America. She was always much engaged with family affairs and family problems, carefully oversaw the education of her daughters and the advancement of her son in his profession, and managed a household with economy and style.

Any consideration of Mrs. Somerville's own cultivation of science must of necessity be biographical and done against the mirror of the social and scientific history of her times. The present account, tracing as it does the development of her scientific interests, skills and career over a quarter of a century, is set out chronologically. The interaction between her individual cultivation of science and the larger cultivation of the subject in the western community — the ideas, opportunities, and limitations imposed on her by the times and so vital to her own participation in the enterprise — are stressed.

Mary Somerville was not a creative scientist in the sense that some of her contemporaries — men such as Wollaston, Herschel or Faraday — were creative scientists. She recognized this fact but knew too that she could make in her own way contributions to the studies that were of paramount importance to her and could bring knowledge of them to a wider public even as she herself satisfied her own inexhaustible thirst for learning. Her life, as related in her personal letters and papers and in those of many of her contemporaries, gives a firsthand account of the cultivation of science during these years, reflecting also the social and political pressures and the rewards encountered by scientists. Her pen delineates, with great fidelity and perceptiveness, the numerous men and women who made up the scientific, literary and political society she knew so well. Their responses to the march of events and to the changes they witnessed, their interactions with each other and with the larger world, and their relations with her and their part in her career emerge vividly from her pages.

After Mary Somerville's death in 1872, she was gradually forgotten by new generations whose limited perspective assigned value only to the science of their own times. Yet Mrs. Somerville's life in science from 1815 through 1840 yields, in many respects, a panoramic view of a period that begins with Waterloo and ends in the year that young Queen Victoria married her Prince Consort. In its particulars her career embodies the attitudes, opinions, the opportunities, the limitations and the many changes characterizing science over this span.

This study examines in some detail the factors that entered into her career, its unfolding, and its influences. Among the broad questions the work addresses are these: How did Mary Somerville move from self-taught provincial to celebrated scientific lady during the period of her London residence? How did she initially enter English and foreign scientific circles? How was she trained and who were her teachers? What actually did she do? Was it significant? Was it useful? Why was her work so well received and she so honoured? Never before have these questions been so fully treated. Their answers are given with the completeness and authenticity the Somerville Collection and other sources provide.

MRS. MARY FAIRFAX SOMERVILLE (aet. 55)

Painted from life in 1835 by Thomas Phillips, R.A., as one in the series
'Eminent Personages' commissioned by the publisher John Murray.
(Reproduced through the courtesy of the Scottish National
Portrait Gallery, Edinburgh.)

CHAPTER 1

SCOTTISH BEGINNINGS

Mrs. Mary Fairfax Greig Somerville, who became 'the queen of [nineteenth-century] science',[1] was born on 26 December 1780 in the manse at Jedburgh, the home of an aunt, Martha Charters Somerville, who later became her mother-in-law. Her own mother, Margaret Charters Fairfax, who barely reached this border town before the baby arrived, was on her way back to Edinburgh from London, where she had seen her naval husband off on a long tour of sea duty. The new mother was very ill after her confinement and the infant was suckled by her sister, Mrs. Somerville, herself lately a mother. This sort of oddity — being born in the house of, and suckled by, a future mother-in-law — is characteristic of the unusual fortune, generally benign, that attended the child throughout her long life.

Her father was an Englishman, Lieutenant William George Fairfax, R.N. (1739–1813), who had gone to sea at the age of nine. The son of an army officer of Bagshot, Surrey, and a grandson of the Yorkshire branch of the family that had produced the great Cromwellian general Sir Thomas Fairfax and the Fairfaxes of the Virginia colony, W. G. Fairfax had long made Scotland his home. He had twice married Scottish wives, both from Burntisland in Fifeshire. It was his second wife, Margaret Charters, who was Mary Fairfax's mother. One of a large family of children of Samuel Charters, Solicitor of Customs for Scotland, she was related to many of the ancient Scottish houses and, like her husband, had a strong sense of blood. Neither of them had any fortune, but birth and position placed them among the gentry.

Mary Fairfax was the fifth of seven children, three of whom died young. Her childhood was spent in Burntisland, a small seaport on the Firth of Forth. There, in a house sold to the Fairfaxes by Samuel Charters and still standing, Margaret Charters Fairfax reared her four young children — the eldest Samuel, then Mary and two younger ones, Margaret and Henry — on her husband's slim navy pay. An indulgent and easygoing mother, she saw no need for any female education beyond learning to read the Bible and newspapers, to write family notes, and to keep household accounts. Her attitude was not unusual; in well-connected Scottish families at this time the sons customarily received excellent educations — attending university and entering the kirk, the legal profession or service in the East India Company — while the daughters mastered social and domestic arts but had only a minimum of book learning.

Not until young Mary's father, returning from a period of sea duty when she was nine, discovered she was hardly able to read, unable to write, and had no knowledge of language or numbers, did this part of her education begin.

He dispatched her, at age ten, to a fashionable and expensive boarding school at Musselburgh — a drastic step for a man of his strong Tory convictions — where for the next twelve months she had the only formal full-time instruction of her life. She emerged from the experience, she recounts in her autobiography, 'like a wild animal escaped out of a cage',[2] but she had developed a taste for reading and acquired some notion of simple arithmetic, a smattering of grammar and French, poor handwriting and abominable spelling.

Over the next years she had occasional lessons in ballroom dancing, pianoforte playing, fine cookery, drawing and painting (under the landscapist Alexander Nasmyth), penmanship, needlework and the use of the terrestrial and celestial globes. A lively and persistent mind, immense curiosity and an eagerness to learn, supported by a robust constitution and quiet, unswerving determination enabled her to take advantage of every opportunity for enlightenment.

At Burntisland, although she had no intellectual guidance, she had freedom to roam the countryside and seashore, observing nature at first hand. She read through the small family library, teaching herself enough Latin for Caesar's *Commentaries*. In Edinburgh during the winter months, family position brought her in contact with educated people in its upper circles and with the rich artistic life and the social gaiety of the Scottish capital at one of its most flourishing periods. A charmingly shy, petite and beautiful young woman — Edinburgh society dubbed her the 'Rose of Jedwood'[3] in reference to her place of birth — she delighted in the parties, visits, balls, theatres, concerts and innocent flirtations that filled the days of popular Edinburgh belles at the turn of the century.

Another and less conventional interest also absorbed her during these years, one that she — in contrast to other scientific women of the nineteenth century — came to without family urging or support and with no example to follow. Between the ages of 13 and 15, a chance glimpse in a ladies' fashion magazine of some strange symbols, said to be 'algebra', aroused her curiosity. Mathematical problems and puzzles were a usual part of such periodicals, inserted for the amusement and entertainment of readers. None of her close relatives or acquaintances could have told her anything of 'algebra', even had she the courage — and she did not — to ask. Her unguided efforts to learn something of this mysterious but strangely attractive subject were fruitless until, overhearing a casual remark by Nasmyth about perspective and Euclid, she was led to persuade her younger brother's tutor, the Rev. Peter Craw — a good-natured young cleric who knew nothing of mathematics — to buy for her copies of Euclid's *Elements* and Bonnycastle's *Algebra*. At the date it would have been unthinkable for a young lady of her position to enter an Edinburgh bookshop and ask for these volumes.[4] After reviewing arithmetic — at which she never became proficient — she began to study these texts on her own. When her father discovered her reading mathematics, he instantly forbade it, fearful that the strain of abstract thought would injure the tender female frame. This view was one widely and long held; Mrs. Somerville herself later believed that her injudicious encouragement of her oldest daughter's intellectual precocity hastened the child's death at age 10. In the late 1790s, Captain Fairfax's strictures against arduous mental effort, together with outspoken criticism of her 'unwomanly behaviour' by aunts and female cousins,[5] drove Mary Fairfax to secret, intermittent application to mathematics but sharpened her resolve to learn the subject. Only her uncle by marriage (and her future father-in-law), the Rev. Thomas Somerville, encouraged her, once he saw her determination to learn. His own four daughters, although better educated than most, had 'little turn for reading'.[6]

Fairfax was an excellent officer — serving twice on the American station, often engaged against the French (who captured him in 1778 and held him on parole for some years) — and advanced steadily in command. In 1797 he was Captain of the 'Venerable', Admiral Duncan's flagship at the Battle of Camperdown. After that great victory, Fairfax, who brought the dispatches to London, was made a knight banneret, given a handsome sword and promised further bounty, which never came. During his years in naval service, he was not often on half-pay, but at a time when many successful naval commanders reaped large profits from their victories, he was singularly ill-rewarded. His children, growing up in genteel poverty, learned to live simply and be self-reliant, honest and truthful, to expect to make their own ways in the world, and to manage what they had to good advantage. They were responsible and clear-headed about money, neither greedy nor miserly. Their fierce Scotch pride drove the family to show a good face to the world, but they avoided extravagance and debt. They believed that Captain Fairfax had been ill-treated but were not embittered by their lack of fortune. From examples they saw all about them and from their own limited experience, they knew how useful a powerful patron could be. They also saw how talent, hard work, determination and a bit of luck could bring rewards of money and position. All these lessons and attitudes shaped Mary Fairfax's character and outlook and later stood her in good stead. Her contemporaries described her as sweet-tempered, cheerful and industrious, ambitious 'to excel in everything she did'.[7]

In 1804 she married a cousin, Samuel Greig, who had come to Scotland from Russia for naval training aboard Captain Fairfax's vessel. Young Captain Greig was son of Admiral Sir Samuel Greig (1735–1788) of the Russian navy and his wife, Sarah Cook, said to be related to the circumnavigator. Sir Samuel himself was a Charters through his mother, Samuel Charters' sister. He and four other young British naval officers had been sent to Russia in 1763 when Catherine II appealed to Britain for aid in reorganizing her navy. The success of that project was due largely to Greig, who remained in the Empress's service until his death. She rewarded him handsomely with promotion, a knighthood, money and large estates in Livonia. His two older sons, Alexander Samuilovich (1775–1845) and Samuel (1778–1807), both found places in the Russian navy and the former also became an admiral.

Mary Fairfax's parents would not agree to her going to Russia, so Samuel Greig left active service to become commissioner of the Russian navy and Russian consul for Britain in London. His bride had never been outside

Scotland and had rarely been away from her family. London was strange, she was lonely, and she missed the vigorous, independent life she had known at Burntisland. Her husband took her to 'his bachelor's house [No. 6 Great Russell Street, Bloomsbury] which was exceedingly small and ill-ventilated'. Decades later she wrote in her autobiography:

. . . I was alone the whole of the day, so I continued my mathematical and other pursuits, but under great disadvantages; for although my husband did not prevent me from studying, I met with no sympathy whatever from him, as he had a very low opinion of the capacity of my sex, and had neither knowledge of nor interest in science of any kind. I took lessons in French, and learnt to speak it so as to be understood. I had no carriage, so went to the nearest church; but, accustomed to our Scotch kirk, I never could sympathise with the coldness and formality of the service of the Church of England . . . I went to . . . [the Italian Opera] as chaperone to Countess Catharine Woronzow [daughter of the Russian ambassador], afterwards Countess of Pembroke, who was godmother to my eldest son. I sometimes spent the evening with her, and occasionally dined at the embassy; but went nowhere else . . . [except to visit] the family of Mr. Thomson Bonar, a rich Russian merchant, who lived in great luxury at a beautiful villa at Chislehurst . . . which . . . [later] became the refuge of the ex-Emperor Napoleon the Third and the Empress Eugenie. . . .[8]

Her first child was born on 29 May 1805, a boy whom they called 'Woronzow' after the Russian Ambassador, Count Simon Woronzow (1744–1832). Another son, William George Greig, arrived before Samuel Greig's death on 23 September 1807. Vice-Admiral Fairfax — promoted the previous year — came to London to help his daughter settle her affairs and escort her and her infants back to the parental roof.

With the newly acquired independence of widowhood and a modestly comfortable inheritance from her husband, Mary Greig set out openly to educate herself in mathematics, ignoring the ridicule and censure of relations and acquaintances who at best thought her foolish and eccentric.[9] The greater part of each day was occupied with her children, and evenings with filial obligations, yet she managed to study 'plane and spherical trigonometry, conic sections and Fergusson's [sic] Astronomy'[10] before attempting Newton's Principia, which she later declared she 'certainly did not understand . . . till I returned to it some time after, when I studied that wonderful work with great assiduity . . . and obtained the loan of what I believe was called the Jesuit's edition, which helped me'.[11]

When she went again into society, she found great help and encouragement for these pursuits from Edinburgh intellectuals. She and another young Scottish widow, Mrs. Jane Kerr Apreece (1780–1855), whose English husband had died within a month of Greig, became great favourites of the elderly Professor John Playfair (1748–1819; F.R.S. 1807), long the leading figure in Edinburgh mathematics and natural philosophy. Playfair liked 'female society and . . . [was flattered by] marked attention from the sex . . . [and] Mrs. Apreece afterwards Lady Davy did her best to captivate him',[12] but Mrs. Greig's dealings with him were mathematical rather than flirtatious. When Playfair learned that she was reading Laplace's Mécanique céleste, he advised her how to get over some of its difficulties. She found him always 'a severe though just critic', his manner 'gravely cheerful, he . . . perfectly amiable and . . . both respected and loved'.[13]

Moreover Mary Greig found support among some of the young Whigs in her social circle, including some of the founders of the successful *Edinburgh Review*, a journal that had long urged widened educational opportunities, including education for women. In this pretty, quiet and liberal-minded young widow — unlike the rest of her family her sympathies were never with the Tories — they saw all the capacity and zest for learning that they asserted for her sex. She in turn admired them for the 'consummate talent . . . [with which] their powerful articles gave a severe and lasting blow to the oppression and illiberality which had hitherto prevailed'.[14] In her memoirs she recalled that she had met at this time 'Henry Brougham who had so remarkable an influence on my future life — his sister had been my early companion' and how she had seen 'the Revd. Sidney [*sic*] Smith, that celebrated wit and able member of the Review at Burntisland where he and his wife came for seabathing'.[15] In time she came to know the other founders of the *Review*, Francis Jeffrey and the two Horner brothers.

Her most helpful mentor at this period, however, was one of Playfair's protégés, the self-educated mathematician William Wallace (1768–1843). Born in Fifeshire in humble circumstances, Wallace had been apprenticed to a bookbinder and had met Playfair while working in an Edinburgh bookshop and attending classes at the university. He had taught himself French, Latin and mathematics, and his abilities so impressed the eminent philosopher that he found the young man a place as assistant mathematical master at Perth Academy. There Wallace remained for nine years, contributing papers on mathematics to the Royal Society of Edinburgh and writing for the *Encyclopaedia Britannica*. In 1803, again at Playfair's urging, he competed in the examination for the post of mathematics master at the Royal Military College at Great Marlow and won. He and Mrs. Greig began a mathematical correspondence, exchanging solutions to problems in the *Mathematical Repository* and discussing difficulties and methods by mail. Wallace was an excellent teacher by post, thorough but encouraging, ready to point out pitfalls and suggest remedies, thoughtful in his approach and clear in his explanations.[16] His own struggles to gain an education made him sympathetic to anyone trying alone to master mathematics. His brother John (d. 1820) was also a mathematics teacher, and when Mrs. Greig began work 'on the higher branches of mathematics and physical astronomy', William Wallace suggested that she engage his brother to read Laplace's *Mécanique céleste* with her. 'I was glad however to have taken this resolution', she later wrote, 'as it gave me confidence in myself and consequently courage to persevere'.[17] In 1811 one of her solutions to a prize problem gained her a silver medal, especially cast, the first of many awards and honours that would come to her. Wallace sent it to his brother, who forwarded it to the winner.[18] It is now at Somerville College, Oxford.

Shortly afterward Mrs. Greig 'put an end to scientific pursuits for a time'.[19]

On 18 May 1812 she married her first cousin, Dr. William Somerville (1771–1860). He had recently returned to Scotland after a long stay in Canada, where he was inspector-general of hospitals and controller of the customs in Quebec. He and his cousin had seen little of each other since childhood, but on meeting again as adults discovered that they had much in common. Both were liberal and tolerant, interested in intellectual matters, ambitious and openminded. Their marriage endured for almost half a century, until William Somerville's death in 1860.

He was the eldest child of the Reverend Dr. Thomas Somerville (1741–1830) of Jedburgh and his wife, Martha Charters. The Somervilles of Cambusnetham were an ancient family but without fortune. Thomas Somerville, whose father had been minister of the kirk at Hawick, received through the patronage of his kinsman, the sixth Lord Somerville, and of his neighbour, the statesman Sir Gilbert Elliot (1722–1777), first the living at Minto and then, in 1772, that at Jedburgh, where he remained for the rest of his life. A man of learning, useful in the business of the church, he knew Hume and other intellectuals of the Scottish Enlightenment. In his early days at Minto he was tutor to Sir Gilbert Elliot's two sons, the elder of whom became the eminent diplomat and statesman, the first Earl of Minto (1751–1814; F.R.S., 1803); the friendship between the two families continued throughout the next century. Thomas Somerville himself won modest fame as the historian of the reign of Queen Anne.

His family — two sons and four daughters — grew up at Jedburgh. The boys from the first were taught that they must find professions; the elder became an army doctor, the younger a writer to the signet. All but one of the daughters married and themselves reared families at Jedburgh. A listing of family events with their dates, compiled by Thomas Somerville and designated his 'Almanac',[20] suggests that William Somerville had some difficulty in fixing upon his path in life. Both sons were sent to Edinburgh to university but William never took a medical degree there. In 1788 he left Jedburgh with his uncle, William Charters, intending to go to Russia to seek appointment as an agent to the Russian fleet. There were many Scots in Russia at that time and opportunities to make fortunes in trade. The two travellers had reached Copenhagen when war broke out and forced them to return home. For a time William acted, rather unsuccessfully, as tutor to the Elliott children at Minto,[21] then in 1794 went to London under the patronage of an important army surgeon, Walter Farquhar (1738–1819),[22] a Scot who had long been in Lord Howe's service.

After some months, Somerville, with the Elliots' help,[23] obtained appointment as a hospital mate.[24] He served in Brazil, then in the forces that were being sent out to take the Cape of Good Hope. He was present when the British seized the colony from the Dutch and remained there for almost eight years. He quickly won the regard of the first governor, General James H. Craig (1748–1812), who appointed him garrison surgeon. After Craig left, Somerville continued to be useful to a succession of Cape governors. He was sent on a variety of missions, including two to unexplored Kaffir country. His short descriptions of the scenery, animals and people, together with the drawings of his companion on one expedition, the artist Samuel Daniell (1775–1811) brought hitherto unknown African fauna and flora to the attention of Europeans. The drawings and notices were published in 1820.[26] A longer account of his journeys in 1800 and 1801–1802[27, 28] was an appendix to John Barrow's *Voyage to Cochin China*.[29] He and Barrow served together for years at the Cape, and both returned to England in 1803.

Sometime around the turn of the century, Somerville had a natural son at the Cape, whom he called James Craig after his patron and whom he acknowledged and educated. When the boy was around 10 or 11, he was brought to Scotland, placed with a family at St. Andrews,[30] then sent to Edinburgh for further schooling. In 1820 James Craig Somerville (c. 1799–1847) graduated M.D. at the University of Edinburgh[31] and, the next year, settled in London, where he became a moderately successful medical man.

William Somerville himself took no formal medical qualifications until 27 June 1800, when he was graduated as a doctor of medicine in the University of Aberdeen. It is unlikely that he did more than send his fees to Aberdeen, since in the summer of 1800 he was on a mission to the interior of South Africa; his official report of that task is dated 14 August 1800.[32] Aberdeen was a popular place for army doctors to acquire an M.D. some years after their actual medical service began; two of Somerville's medical patrons took their medical degrees there — Sir Walter Farquhar in 1796 and Sir James McGrigor (later head of the army medical department) in 1804.

After Somerville's return to England in 1803, he appears to have been for a time in Scotland, but by early 1804 he was again in London, hoping to establish himself as a physician and serving meanwhile as assistant to the Inspector General of Army Hospitals.[33] His stays in Scotland tended to be short; there was little employment for him in Jedburgh and his liberal views on politics and other matters were anathema to most of his conservative relatives. In London he seems to have attracted little attention and apparently knew none of the London scientific men.

In 1805 when General Sir James Craig was given command of a large force sent secretly against Bonaparte in the Mediterranean, he put Somerville in charge of his medical staff. After Craig's successful occupation of Naples and Sicily, however, his health failed and he returned to England with Somerville in attendance. The doctor seems to have been on half-pay when he visited his parents in July 1806. On 2 September he married Anne Rutherford, the heiress of Knowsouth, near Jedburgh, and shortly afterward they sailed for Malta, where Somerville was stationed for more than a year. There their son was born in 1807 but lived only a few months.[34] Mrs. Somerville herself died at Falmouth early in 1808, attended in her last illness by her husband, his brother and his sister.[35]

Craig the previous year had been made governor-general of Canada and Somerville joined him in Quebec. During his stay in Canada the doctor acquired large tracts of land in the province and had additional income from his controllership of customs. One of his fellow-officers was his cousin Henry

Fairfax (1790–1860), younger brother of Mrs. Mary Greig. When Craig
became ill again in 1811, Somerville again accompanied him back to England.
He was in attendance on the general when that distinguished soldier died in
January 1812.

On a short visit to Scotland in December 1811, William Somerville saw his
cousin Mary for the first time in some years. They now had much in common.
Both had lost spouses, both had sons to rear. Their views on politics, religion,
on social reforms and on many other questions were compatible and much
more liberal than those of the majority of Scots. Of this cosmopolitan medical
man, Mary Greig wrote:

. . . he generally lived in London so that he was seldom with his family with whom he
was not a favourite on account of his liberal principles, the very circumstance that was
an attraction to me. He had lived in the world and had gentlemanly manners, spoke
good English and was emancipated from Scotch prejudice.[36]

He was also stout in defence of her independence and of her studies, seeing
nothing improper in her pursuit of mathematics and astronomy. Her own
parents had come to accept this aspect of her conduct, but among the rest of
her relations, only William Somerville and his father approved. She had
refused several proposals of marriage since returning to Edinburgh[37] but
accepted her cousin's. Their parents were delighted at the match.

William Somerville was 41 years old when he wed for the second time. He
had spent the greater part of the past 17 years away from England and
Scotland and was ready to settle down and rear a family. When an appointment as deputy inspector of hospitals at Portsmouth was offered him, he
accepted and pressed for an early wedding. His bride was not yet 32 and,
aside from three unhappy years in London, wholly provincial. She was,
however, charming, pretty, gifted, intelligent, ladylike and observant. On
their way southward they spent some time in the Late District. Nearing
London, they stopped at Great Marlow to see Mrs. Somerville's mentor,
Professor William Wallace. He took them to call on William Herschel
(1738–1822; F.R.S., 1781) at nearby Slough. Wallace gave Mary Somerville
Herschel's reply to his note asking permission to bring the Scottish couple to
see the great telescope. In it the eminent astronomer declared he would be
'very happy to see the Lady and any of the party you mention . . . and you
may be assured that the *trait* in the Character of the Lady to be a good
mathematician without *Wrangleship* will be highly esteemed'.[38] This memor-
able visit was the couple's first taste of English scientific society.

They soon returned to Edinburgh, for William Somerville in 1813 became
head of the Army Medical Department in North Britain.[39] There, on 5 March
1813 their first child, a daughter whom they called Margaret Farquhar, was
born. Motherhood, however, did not deter Mrs. Somerville's mathematical
studies. Urged on by an approving husband, she purchased, at Wallace's
direction, a small, splendid library of mathematical works, mostly French. In
her autobiography she carefully lists these works:

. . . Francoeur's pure mathematics and his elements of Mechanics, La Croix's algebra and his large work on the Differential and Integral Calculus together with his work on finite differences and series, Biot's analytical geometry and astronomy, Poisson's Treatise on Mechanics, Lagrange's theory of analytical functions, Euler's algebra, Euler's isoperimetrical Problems (in Latin), Clairault's figure of the earth, Monge's application of analysis to geometry, Callet's logarithms, LaPlace's Mécanique Céleste and his analytical theory of probabilities &c &c.[40]

Apart from the Euler in Latin (1740), all these books were published after 1795, the majority of them dating from between 1805 and 1810 and the most recent being the Poisson (1811). The collection, then, was up-to-date. All these volumes — and two others, Lagrange's *Traité de la résolution des équations* (1808) and Legendre's *Éléments de géométrie* (1809) — are among the books belonging to Mrs. Somerville presented by her daughters after her death to the new ladies college at Hitchin (now Girton College, Cambridge).

Many years later Mrs. Somerville wrote:

I was thirty-three years of age when I bought this excellent little library. I could hardly believe that I possessed such a treasure when I looked back on the day that I first saw the mysterious word algebra, and the long course of years in which I had persevered, almost without hope; it taught me never to despair.[41]

Her mastery of this difficult and often quite new mathematical material contrasts sharply with the ignorance prevailing amongst most male mathematicians in England at the time. Scotland had for some decades been mathematically more venturesome than its southern neighbour and was, for a number of reasons, more receptive to the new French analysis. Ancient cultural ties and sympathies existed between the two peoples, and there were more recent links, arising from the influx of French prisoners-of-war held in North Britain during the years of trouble with France. Some of these men were graduates of the École Polytechnique or of other French technical schools. They were well trained in French mathematics and often willing to teach it to interested Scots. It was Mary Somerville's skills at French analysis that made it possible for her, nearly two decades later, to bring this branch of mathematics to English readers.

William Somerville insisted that his wife broaden her studies. The young man, Donald Finlayson, acting as his secretary and as tutor to her son Woronzow was 'a remarkably good Greek scholar',[42] so she read Homer with him an hour each morning. His successor, with whom she also had lessons, was an able botanist, his brother George Finlayson (1790–1823), who later won a reputation as a naturalist on expeditions to southeast Asia. Both Somervilles embarked 'with zeal' on the study of mineralogy, the popular science at the moment. They made the acquaintance of Professor Robert Jameson (1774–1854; F.R.S., 1826), who taught them the Wernerian system he favoured. They met Professor John Leslie (1776–1832) and liked him, finding him 'a man of original genius, full of information on a variety of subjects, agreeable and good natured'.[43] Mary Somerville also encountered for the first time David Brewster (1781–1868; F.R.S., 1815). The son of the Jedburgh schoolmaster, this scientist had been known to William Somerville from childhood but his path and Mary Somerville's had not previously crossed.

William Somerville began to move in Edinburgh scientific circles, not as a scientific practitioner but as a genial friend of science, the same role he would find in London. He became a Fellow of the Royal Society of Edinburgh in 1813[44] and joined Jameson's Wernerian Society. An affable companion and a travelled and cultivated gentleman, he was quickly at home in this company. His position with the Army medical department gave him easy entry into medical and scientific circles.

His wife, long accustomed to Edinburgh society, was busy now with her studies and her young family. Her two sons by her first marriage lived with them and attended Edinburgh schools. The younger boy died in March 1814, aged nine.[45] His mother was pregnant at the time with her second child by Somerville, a son born the following July. This child died too, in February 1815.[46] Nine months later, on 15 November 1815, a second daughter, Martha Charters Somerville, was born.[47]

The year of Waterloo was a difficult one for the family, beset as they were by these sorrows and strains. William Somerville, on learning of the fighting in Belgium, left Edinburgh on 30 June with his good friend, John Thomson (1765–1846), professor of military surgery at Edinburgh University. They were anxious to be useful and eager to learn as much as possible from firsthand observation of the wounded. In the next weeks they visited hospitals in Antwerp,[48] Brussels,[49] and Ghent and then went on to Paris, where they witnessed some of the historic events of the weeks following Napoleon's defeat. By the end of August William Somerville was back in Edinburgh,[50] comforting his anxious wife who had been poorly for some months.

Toward the end of the year, circumstances improved. Their new daughter arrived in November; the older one was flourishing. And from London came splendid news: on 28 December 1815 William Somerville was named to a new office. He was made one of the two Principal Inspectors of the Army Medical Board[51] and ordered to take up the post in London under the Director General, a fellow-Scotsman and friend, Sir James McGrigor (1771–1858; F.R.S., 1816).

CHAPTER 2

LONDON BEGINNINGS

The Mary Somerville who returned to London in 1816 was a far different person, in far different circumstances, from the shy and lonely young widow who had left the capital in 1807 eager to go back to Scotland with her infants and to settle again under a parental roof. During five years of widowhood and through her remarriage she had gained a new independence and self-confidence. Although still timid and provincial, she found adjustment easier in 1816 than it had been in 1804. William Somerville — ever buoyant about the future, always 'partial to London'[1] — introduced her to a London life much more to her liking than that she had known as the bride of Samuel Greig. Unintentionally Greig had been narrow and neglectful. Somerville, on the other hand, was instinctively liberal and encouraging. His own talents in mathematics and science did not rise to any level of mastery, but he appreciated the fact that hers did. Ambitious and gregarious, he perceived his talented wife as a distinct asset, a congenial and useful helpmeet who, through her accomplishments, could assist his rise in London as she had done in Edinburgh.

Late in January Dr. Somerville came to London to make arrangements for moving his family. He took a house in Bloomsbury, the part of the city his wife knew best. Before returning to Scotland, he enrolled as an annual subscriber to the lectures at the Royal Institution.[2] His intention that they should continue their scientific pursuits is clear.

The couple and their two daughters left Edinburgh on 15 March 1816 and by mid-April were settled into their London residence at No. 12 Queen Square. The two boys, Woronzow Greig and James Craig Somerville, remained behind to continue their schooling. The Somervilles' London acquaintance was small: an elderly spinster aunt, sister of Thomas Somerville, who had lived there for many years; a merchant family called Rucker, also known to the Jedburgh Somervilles; professional colleagues of William Somerville; and a few of the numerous Scots who made their way south on various missions or had settled in the English capital. The Scottish colony in London was diverse and influential, ready to push for the advancement of their countrymen and important to the Somervilles' future.

Through a fellow-Scot in Edinburgh, Leonard Horner (1785–1864; F.R.S., 1813), the Somervilles became acquainted with the couple who acted as their chief sponsors in their early London months and who first introduced them to London scientific society. On 14 March 1816 Horner wrote the London physician and chemist, Alexander J. G. Marcet (1770–1822; F.R.S., 1815), a friend with whom he was in frequent correspondence:

You have probably heard of Dr. Somervilles appointment to the Army Medical Board — He is a very good fellow, & his wife a very interesting woman. She is a person of very extraordinary acquirements, particularly in mathematics. But she has not a shade of blue in her stockings. They will be a great loss to the Society of Edinburgh — Mrs. Horner & I have hardly yet formed an acquaintance. They have taken a house in Queen Square and set out for London tomorrow. She is desirous of being acquainted with Mrs. Marcet & my sisters have requested me to ask the favour of Mrs. Marcet to call on her when she gets to town, which I do with pleasure as she is one that excites great interest among all who know her. She will send her card with a note from my sister to Mrs. Marcet when she reaches London.[3]

At the time the Marcets lived at No. 23 Russell Square, a few hundred yards from the Somervilles' new residence. Jane Marcet was prompt in calling. A month later her husband reported to Horner, 'Mrs. Marcet has seen Mrs. Somerville, and likes her very much; she will be a great acquisition in our neighbourhood'.[4] Thus began a friendship between two scientific ladies that would end only in death. Through the Marcets the Somervilles were introduced into the most distinguished scientific circle in London, one that encompassed W. H. Wollaston, Henry Kater, William Blake, Henry Warburton and others.

Alexander and Jane Marcet had for nearly two decades offered warm hospitality not only to the Anglo-Swiss community in London but to many foreign visitors and to literary and scientific society in general. As a young man Marcet had been exiled from his native Geneva during its 'reign of terror' in 1794.[5] He fled to Edinburgh, where in 1799 he took his medical degree, then settled in London. In 1799 he married Jane Haldimand (1769–1858), daughter of a wealthy Swiss banker and his English wife. Marcet became a naturalized British citizen and in 1803, physician and chemical lecturer at Guy's Hospital. An able experimentalist, he carried out over the next years important chemical and medical researches and actively collaborated with J. J. Berzelius. One of the founders of the Medical and Chirurgical Society in 1805, Marcet was also an early member (1808) of the Geological Society and a Fellow of the Royal Society. He and Wollaston were good friends, and it was to Wollaston that he dedicated his 1817 monograph on calculus disorders.[6]

A decade earlier Marcet had persuaded his wife to undertake a simple book on chemistry written in such a way that young ladies captivated by Humphry Davy's popular lectures at the new Royal Institution could 'understand as well as admire him'.[7] Jane Marcet's Conversations on Chemistry,[8] published anonymously in 1806 soon became — and long remained — 'the general textbook [on chemistry] for young men in Great Britain and the United States',[9] gaining a special immortality by awakening in 1810 a young bookbinder's apprentice, Michael Faraday, to the subject.[10] She continued her authorship with a series of popular 'Conversations', which served as useful introductions to various studies. Jane Marcet was a 'sensible . . . unaffected . . . [rather] plain [woman] . . . under the strongest conventional influences'[11] and with no pretension to scientific knowledge.

An excellent hostess and helpful friend, she was warmly regarded by London intellectual circles for her own virtues and as an expositor of simple science.

The hospitable reception the Marcets extended to the Somervilles in the spring of 1816 was all the more welcome in view of the calamity overtaking the newcomers almost as soon as they reached London. In March 1816 after the Army estimates were presented to the House of Commons, the Aberdeen radical Joseph Hume (1777–1856; F.R.S., 1818) launched the first of what became an annual campaign for drastic reduction in military spending. His demands had strong popular support. Waterloo had removed the French threat, but severe domestic unrest arising from unemployment, new taxes, the Corn Laws, and the festering problem of workers' rights gripped the country. The government was forced into an appearance of retrenchment; Lord Palmerston, the Secretary of War, made the prudent gesture of reducing what he termed 'unnecessary expenditures'[12] chiefly by lowering salaries and abolishing new posts.

William Somerville was one of the first victims of Palmerston's economies. Leonard Horner wrote his friend Marcet on 24 April:

. . . Since they [the Somervilles] went to London his appointment has been taken away from him, we hear; one of the instances in which ministers have redeemed the Regent's pledge of economy — very justly too, I think — it will not be long before they must come to the reduction of some of the large items of their profligate expenditure. . . .[13]

The high hopes with which Somerville and his family had arrived in London were thus quickly blighted. Furthermore, moving house had put them considerably out of pocket. The resiliency and optimism which William and Mary Somerville displayed again and again in their long life together were soon apparent as they cast about for some solution to their difficulties. In June Mary Somerville wrote her father-in-law:

. . . We go on very well in our new station . . . tho I did not like it at first I now find it [London] very agreeable, the house we are in is comfortable but distant from every thing, and had we been to remain in London we must have changed it. however as matters stand we are fortunate in having been saved a great expense — Nothing new has transpired regarding the [Army] Board, but it is believed to be certain that William at least must go on half pay. we are both delighted with the prospect, and have fixed to spend the winter in Italy. Sir W. Farquhar and William's friends here strongly recommended this plan as being more conducive to the advancement of his views than any other, as wherever we may be Sir W. would introduce him as Physician to the English families abroad by which means he would get into good practice that might be continued when we come home. at all events it is spending our time very pleasantly. . . .[14]

Their intention to go abroad had been made known even earlier in Edinburgh; Robert Jameson mentions the plan in a letter written in mid-

May to William Somerville.[15] But the trip had to be postponed, for by the end of June Mary Somerville knew that she was again pregnant. Her last child, a daughter called Mary Charlotte, was born in London in early 1817. The final stages of the pregnancy were so difficult that she was for some time confined indoors, but her recovery after delivery was rapid and the child was 'stout and . . . fine'.[16]

Despite the fact that the Somervilles were '. . . in utter darkness as to . . . future plans',[17] the months they passed in London in 1816–17 were neither tedious nor wasted. William Somerville's natural ebulliency, his tenacity in the face of adversity, and encouraging responses to his continued complaints and exertions — along with his wife's pregnancy — persuaded him to remain a little longer as he was. Disappearance of his new post did not wholly sever his ties to the Army Medical Department, for early in November 1816 he made a trip to Cornwall to inspect medical establishments there.[18] The occasional private patient came his way and he continued to hope for a favourable resolution to his situation even though the 1817 Army lists did not carry his name.

Somerville's unexpected leisure gave him time for new friends and new affiliations, both of which would prove important to his wife's future — but then undreamed of — career. To his subscription to the Royal Institution he soon added membership in the Geological Society. Alexander Marcet and the two Horner brothers were already fellows, as were a number of the Somervilles' Scottish friends, including Robert Ferguson of Raith, Professor John Playfair, Professor Robert Jameson, Sir James Hall, Charles Bell, Lord Webb Seymour, George Birkbeck, and John Rennie. J. A. Rucker and D. H. Rucker were also members. Somerville's election was prompt; he is listed sixth among the 21 ordinary members of 1816.[19] Anxious to establish itself and to encourage support for its specialty, the Geological Society welcomed new fellows and attracted not only serious practicing geologists and mineralogists but many less informed men mindful of the prestige to be gained through endorsing 'modern science' and eager to enter London scientific circles and enjoy London scientific company.

Somerville became a fellow also of the Linnean Society,[20] which too had a large Scottish contingent. On 24 April he was proposed for fellowship in the Royal Society of London. His certificate — which appears to be written in the hand of Thomas Young — reads:

William Somerville M.D. of Queen's Square Bloomsbury, F.R.S.E. F.L.S. F. Geol. Soc., a gentleman well acquainted with the sciences in general, and with natural history and mineralogy in particular, being desirous of becoming a Fellow of the Royal Society, we recommend him from our personal knowledge as highly deserving of the honour he solicits, and likely to be a valuable member of the Society.

Seventeen signatures follow in this order: Thomas Young, W. H. Wollaston, Alexander Marcet, Charles Konig, Humphry Davy, Richard Sharp, Gilbert Blane, J. F. W. Herschel, P. M. Roget, Henry Holland, James McGrigor,

John Yelloly, Robert Brown, William Blake, B. C. Brodie, John Pond and Wilson Lowry.[21]

The list not only attests to the range and distinction of the Somervilles' earliest London scientific acquaintance but exhibits the strands that would make up their scientific future — one Scottish, one medical, a third military, and a fourth metropolitan. It suggests also the ready and useful bonds created through shared memberships in several scientific bodies. Brown, McGrigor and Blane were Scotch born; Holland, Marcet, Roget, Yelloly and Young had studied at the University of Edinburgh and had strong Scottish ties. Nine of the seventeen signers — Blane, Brodie, Holland, McGrigor, Marcet, Roget, Wollaston, Yelloly and Young — had qualified as medical men, and four of them — Brodie, Holland, Roget, and Yelloly — had substantial London practices. Blane was the chief medical officer of the navy, McGrigor since 1815 of the Army. Seven — Blake (then P.G.S), Davy, Holland, Lowry, Marcet, Wollaston, and Yelloly — were, like Somerville, members of the Geological Society, while three — Brown, Konig and Young — were enrolled with him in the Linnean Society. Seven of the signers — Davy, Wollaston, Young, John Pond (the Astronomer Royal), Robert Brown, John Herschel (who was working closely with Wollaston) and Marcet — were widely hailed for their experimental researches. Sharp — called 'Conversation' from the brilliance of his discourse — was part of London literary and political society and a reform M.P. Lowry was a well-known engraver and inventor; his wife, a mineralogist, was giving Mary Somerville lessons in the subject at the time. Charles Konig, a German-born and German-educated mineralogist, had since 1813 been Keeper of the Department of Natural History at the British Museum.

Twelve of the 17 signers had become Fellows of the Royal Society within the decade and a high proportion of them were active in the management of its affairs. Wollaston and four others — Young, Roget, Konig and Herschel — were at various times secretaries. Wollaston, Davy and Brodie would each become its President. In short, the array of talent and influence assembled in 1817 in support of the election of a relatively obscure Scottish army doctor was remarkable at any time and hints at the potency and vitality that the Somervilles' scientific associations would have in the 1820s. A certificate required but three signatures; endorsement by seventeen fellows represented strong but not unprecedented support.

Thus within a little more than a year of their arrival in London and despite the ill fortune of losing his Army place, Somerville had associated himself with three of the four leading scientific societies of the capital, and he and his wife had established ties with the best London scientific society. The ease with which this couple — without significant scientific reputations or training, without fortune or noble patronage but intelligent, personable, well-introduced and genuinely interested in science — made their way into a congenial group that encompassed some of the most gifted active scientists of the day demonstrates how strong was the bond of science, how acceptable was a dilettante concern with the subject, how important a compatible personality

and evident enthusiasm. For those less well-introduced or less socially acceptable than the Somervilles, the way was often more toilsome but the end was
still attainable: Michael Faraday in his rise from bookbinder's apprentice to
London chemical lecturer and F.R.S. in these same years showed how ability
and determination could triumph over origin and adversity.

Nor did the Somervilles lose touch with their Edinburgh friends. Mary
Somerville and Jameson kept up a mineralogical correspondence. He sent her
a copy[22] of his new edition,[23] she made him a present of ore samples.
Jameson's letter of thanks suggests one of the reasons why serious scientists
regarded Mary Somerville so highly:

. . . I am delighted with the rapid progress you have made both in the knowledge of
minerals & the arrangement & extension of your cabinet. It is enchanting to me to
have met with one so highly gifted still zealous and ardent in the study of my favorite
pursuit — Many indeed begin but few continue to advance — novelty as it is called
being their object & that being gratified some other amusement must be sought
after. . . .[24]

William Wallace in Marlow continued to give mathematical help; two of
the three letters from Wallace among the Somerville papers date from this
period and both illustrate his method of written instruction, his advice on
mathematical reading, and his emphasis on French mathematics.[25]

Mary Somerville found that she could readily combine continued study
with London domestic and social demands. As soon as the family was settled
in Queen Square, she resumed work in mathematics and mineralogy. During
the months of her pregnancy, she spent a good deal of time in these studies
and in painting. Lessons in mineralogy from Mrs. Wilson Lowry and in
painting from the landscapist John Glover (1767–1849) improved her skills.[26]

In the summer of 1816 two French scientists, D. F. J. Arago (1786–1853;
F.R.S., 1818) and J. L. Gay-Lussac (1778–1850; F.R.S., 1815) were in England for a time and much in London scientific society. The Somervilles
became acquainted with both. In the following spring another Frenchman,
J. B. Biot (1774–1862; F.R.S., 1815), arrived to complete the French measurement of the arc of the meridian, a project on which he and Arago had been
uneasily collaborating for some time. The Somervilles met him — at the
Henry Katers according to notes Mrs. Somerville made later for her autobiography.[27] Biot had been told in advance of her interest in mathematics and
astronomy and found her a charming curiosity. She later reported that he
confessed to her, 'I expected to see an old woman and was surprised to see a
young and pretty one'.[28] This anecdote was repeated again and again by
others as her fame grew.

These French savants were often at the Marcets and grew to know all the
circle that gathered there so frequently. When Biot moved on to Scotland to
undertake his measurements, he carried letters of introduction from William
Somerville to various distinguished Edinburgh residents, all of which proved
useful. His long letter of thanks to the doctor[29] — filled with expressions of

admiration for Mrs. Somerville and stressing his eagerness to make her acquainted with Madame Biot — makes it clear that by early 1817 the Somervilles, the Marcets, the Katers, the Thomas Youngs, Wollaston, Henry Warburton (whose brother-in-law General Sir Howard Elphinstone [1773–1846] Somerville had known at the Cape), William Blake and Sir John Saunders Sebright had already settled into a close and easy association that was congenial also to visiting French savants.

Biot, having requested and received from Dr. Somerville permission to write Mrs. Somerville, sent her a lively and lengthy account of his life in Scotland, his work on the meridian measurement and some instructions about her coming visit to Paris, as well as his thanks for the kind reception he had met in London.

Pour vous, Madame, vous allez à Paris; et vous y trouverez je vous assure, des personnes qui ont déjà une très grande envie de vous voir. Madame Biot n'a pas besoin que je lui écrive une nouvelle lettre pour lui faire connaître Madame Somerville. Il suffit Madame Somerville veuille bien lui apprendre son arrivée, en lui disant *je suis Madame Somerville,* et je vous assure que ce nom-là fera un meilleur effet que tout mon griffonnage. N'oubliez pas surtout de dire à Madame Biot de faire connaître à Monsieur Somerville Monsieur Brongniart, le directeur de la fabrique royale de porcelaines de Sèvres, qui a employé et fait exécuter en émail les beaux dessins du voyage du Cap. Et pour tout ce qui pourra vous être commode ou agréable, soyez sûre que vous trouverez Madame Biot aussi empressée à le faire que vous et Monsieur Somerville l'avez été pour son mari . . . J'espère bien cependant, Madame, avoir le bonheur de vous voir encore à Paris avant votre départ pour Londres; j'espère vous y remercier ainsi que Monsieur Somerville de toutes vos bontés. Veuillez présenter mes respects et mes souvenirs à ceux de vos nombreux amis qui m'ont accueilli si obligeamment à Londres: Monsieur et Madame Young, Monsieur et Madame Kater, le Docteur Wollaston, Monsieur Blake, Sir John Sebright, et bien d'autres encore qui feraient une trop longue liste pour cette feuille de papier. . . .[30]

Plans for going abroad had been revived after Mary Somerville's confinement and recovery, but the Somervilles' principal intention now was a visit to Paris.[31] Further travels would depend on Dr. Somerville's situation at home. By July their preparations were made. They gave up the house in Queen Square and the two Somerville infants — Martha, twenty months old and Mary, not yet six months — were placed in the care of William's childless married sister, Mrs. Janet Pringle of Ferney Green, near Kendal, with whom they would remain during the long period that their parents did in fact stay on the Continent. The two boys, Woronzow Greig and James Somerville, were left in the care of relatives in Scotland.

CHAPTER 3

THE FIRST TRIP ABROAD

1. Paris and its Scientific Society, 1817

On 17 July 1817 Dr. and Mrs. Somerville, their four-and-half-year old daughter and her maid, and Somerville's only brother, Samuel Charters Somerville (1776–1823), Writer to the Signet in Edinburgh, sailed by packet from Dover to Calais. It was Mary Somerville's first trip abroad and it would leave her, bad sailor though she always was, with an incurable taste for travel. In the initial weeks of their travels she kept a journal,[1] faithfully recording their activities in some detail as well as noting her impressions of sights and people, snatches of conversations, judgements and opinions, and much factual information. It clearly conveys her astonishment and delight — occasionally shock and disapproval — at foreign scenes and foreign customs. Hers is a fresh, acute and observing eye, and the diary gives one of the most detailed accounts yet available of the reception of British visitors by the Parisian scientific community in 1817. Entries cease after a few weeks, but a number of letters written by the travellers and some rough notes made by Mrs. Somerville remain in the Somerville Collection to continue an account of their journey and to illustrate her capacity to grow, adapt and change. The Mary Somerville who left Dover in mid-summer 1817 was intelligent but provincial, satisfied with things Scottish or English. The woman who returned to London in 1818 was far more cultivated and cosmopolitan.

France in July 1817 still showed many scars of the Napoleonic defeat. Wounded and discharged soldiers in tattered uniforms and cocked hats were among the hordes of beggars everywhere. The rubble of war and evidences of shelling littered the towns and battle sites. Poorly tended fields and animals and awkward farm equipment accounted for the manifest poverty in the country districts. All were remarked in Mrs. Somerville's journal, which displays also three characteristics of her later scientific writings: (i) a passion for quantification; (ii) a passion for nature; and (iii) a passion for technology. The diary shows too an abiding interest in art and architecture and a taste that — though far less informed than it would be by the end of the tour — always favoured simplicity, proportion and rich colour.

The party arrived in Paris on the evening of Monday, 21 July, established themselves in a hotel on the Rue de la Paix, then immediately began a round of sightseeing and calls that would continue until their departure a fortnight later. William Somerville, who had spent some years in Quebec, spoke and read French easily, his wife more slowly and timidly. Somerville had last been in Paris in the summer of 1815, after Waterloo; he had some familiarity with the city and was an experienced tourist.

Their reception by the French scientific community — and during their stay the greater part of their time was spent in scientific company — was the most singular aspect of their brief stay in the French capital. The Somervilles were received in much the same fashion as were notable visiting philosophers. Even during the Napoleonic wars, English scientists had been welcomed to Paris by their French counterparts; Davy, for example, made his first trip abroad in 1813 and was cordially accepted.[2] From his account and from other sources[3, 4] the kind of hospitality extended by French savants to visiting English philosophers is evident.

Customarily the foreigner was received by savants he already knew, either through previous meetings or correspondence, and by those to whom he brought letters of introduction. They in turn entertained him in their laboratories or museums, took him sightseeing, and — at evening gatherings or during the day — presented him to their Parisian colleagues and their wives and to other visitors. The newcomer would be invited to attend at least one session of the Institut de France and of the Académie des Sciences. He would visit the Muséum d'Histoire Naturelle, the Jardin des Plantes, and the Paris Observatory, being shown over each — if he were sufficiently distinguished or highly favoured — by its director or a designated assistant. Should the guest wish to see additional sights or meet non-scientific personages, arrangements to do so would be made by his French hosts. There was generally a tour of Paris that included visits to the Louvre, the gardens and palace of the Tuileries, Notre Dame, and other buildings and monuments of interest. In all likelihood he would be invited to spend a day at nearby Arcueil, where the two premier French scientists, Laplace and Berthollet, had adjoining country estates and where their scientific circle — the informal but powerful Society of Arcueil — gathered to talk science, politics, and literature and to enjoy country pleasures and the company of renowned guests. The pattern of entertainment afforded the visiting savant in Paris was one of continuous and cordial hospitality interwoven with sightseeing and scientific intercourse. It changed little through the early decades of the century.

Thanks to their introduction in London to Arago and Gay-Lussac and to Biot's letters to his wife about them and to her exertions on their behalf, the Somervilles were treated as if they were practicing scientists of distinction. Letters of introduction also opened friendly doors among the Scottish and English colonies in Paris, as well as among some non-scientific French. At the end of their stay, Mary Somerville could justifiably state to her mother:

. . . we continued to like . . . [Paris] to the last and had every reason to do so from the uncommon attention we met with, besides seeing all kinds of public institutions we were invited with much hospitality to the houses of the most celebrated philosophers of France, who made entertainments on purpose to make us acquainted with every one who was eminent in science. . . .[5]

Their first visit was not to Madame Biot, as her husband had instructed, but to deliver a letter of introduction to M. Jean André Henry Lucas (1780–1825), keeper of the Cabinet of Mineralogy at the Muséum d'Histoire Naturelle there. Lucas, the son and grandson of functionaries of science,[6] escorted them through the Muséum, explaining its organization and function. He then conducted them to the apartment of the Abbé Rene Haüy (1742–1822; F.R.S., 1818), who had been made a professor there by Napoleon. This great crystallographer, Mrs. Somerville noted,

> . . . took much pains to make our visit interesting, explained the plan of his work, shewed us models in wood for the illustration of his Theory, and likewise many drawers of minerals . . . [and after exhibiting and explaining a number of specimens] gave us a few crystals of quartz and a copy of his traite [sic] de ——— to Somerville.[7]

The title which Mrs. Somerville omitted — one has the impression reading the diary that she expected to fill in the blanks later — may well have been Haüy's *Traité des Pierres Précieuses* published in 1817. Mineralogy was at the time the Somervilles' chief science, and their interest in Haüy's views was indicative of a growing scientific sophistication. His notions were opposed to those of A. G. Werner (1750–1817), whose theories were supported by their Edinburgh mentor Robert Jameson.

From the Muséum they went to the Paris Observatory, where they saw the astronomer and mathematician C. L. Mathieu (b. 1784), Secretary of the French Board of Longitude since 1806 and Biot's collaborator in some pendulum and meridian measurements. Mathieu's wife was Arago's sister, and the two families resided at the Observatory. Another sister[8] was married to A. T. Petit (1791–1820), who with Pierre Dulong (1785–1838) enunciated the law of atomic heats. While at the Observatory the Scottish couple called on Arago and Gay-Lussac, but both were out.

Finally, they went to the 'Collège Royal' — the Collège de France — where Biot had become professor of physics and mathematics in 1800 (he would hold the post until his death in 1862) and where his wife, as promised, received them warmly. Over the next two weeks Madame Biot undertook to make them acquainted with all the scientific personages in the French capital. She herself had a strong penchant for science and had assisted her husband for a number of years. Biot had set about her education after their marriage so that she might 'grace the company of his scientific friends'[9] and found her an apt pupil. A woman about Mary Somerville's age, she was the sister of one of his classmates at the École Polytechnique. Born Françoise Gabrielle Brisson, daughter of a distinguished physician and natural philosopher of Beauvais, she had married Biot, then 23, when she was 16. Madame Biot was a good linguist, reading English, German and Italian, and had translated E. G. Fischer's *Lehrbuch der Mecanischen Naturlehre*. Biot had published this translation under his own name,[10] although declaring in its preface that he knew no German and that the book had been done by a 'person very dear' to him. Their only son, Edouard Constant Biot, later a famed Chinese scholar, was about the same age as Mrs. Somerville's son Woronzow. Biot's ready acceptance of Mary Somerville arose no doubt from similarities he discerned between her interests and those of his wife. Of the Frenchwomen she met in 1817, Mrs. Somerville found Madame Biot the most companionable.

Of these ladies she wrote, five days after arriving in Paris:

. . . It is by no means a fair thing to give an opinion of any set of people on a short acquaintance, yet I could not help being struck by the difference between the accomplishments of the french and English Ladies. Among all I have met only one pretended to know a little music and that was poor indeed, two drew a little, in language and science I met with none except Mme Biot. . . . Certainly I found none of that high cultivation of mind and elegance of manners so constantly seen in England, not among the higher classes alone but widely diffused throughout the nation. Dress is a great object among the French ladies and forms a frequent subject of conversation.[11]

The Somervilles' first jaunt with Madame Biot was to the Louvre, where they spent the morning of 24 July 1817.[12] On the following day they again visited this great museum, this time with Arago and Gay-Lussac, then joined Madame Biot at the Institut de Paris.[13] She showed them its library adorned with a statue of Voltaire and bust of La Grange, its private meeting rooms and its public theatre. Mrs. Somerville describes in detail the uniforms of members. They witnessed a long, tedious session — from 3 to 5 p.m. — devoted to a series of talks of monumental dullness. Mrs. Somerville comments in her diary that she '. . . saw no spectator so tired as the members themselves seemed to be'.[14] The Scottish visitors were happy to leave the world of organized scholarship for an evening at the theatre.

Biot had suggested a visit to Sèvres and provided a letter of introduction to the Director of the Porcelain Works, the savant Alexandre Brongniart (1770–1847; F.R.S., 1825). Brongniart had used some of William Somerville's descriptions of South African animals in designing porcelain figurines and they were anxious to see them. Unfortunately he was absent when they arrived on 26 July, but they were given an interesting tour of the manufactory and saw the African animal replicas.[15] It was a busy day for the visitors, for on their way to Sèvres they stopped at St. Cloud and in the evening dined with Madame Biot. Of this occasion Mrs. Somerville wrote, '. . . stile of dinner singular but very good . . . a great deal of conversation about the 100 days — the flight of B[onaparte] and return of the King'.[16]

In the evening of the 26th they went to the Cuviers',[17] who lived at the Jardin des Plantes, where Cuvier (1769–1832; F.R.S., 1806) had been professor since 1802. In one of the few character sketches in her diary, Mrs. Somerville describes the celebrated anatomist and geologist as a man with '. . . a very singular turn of countenance, not handsome but agreeable with a very genteel manner great point in his conversation, and modesty'.[18] Arriving a bit early, the Somervilles walked in the gardens until the party assembled; only two of its members — Madame Biot and 'a Venetian lady of acquirements', the Countess Albrizzi (1760–1836) — are named.[19] Conversation during the evening dealt with papers read that morning at the Institut, then moved on to a condemnation of Lady Morgan's recently published work on France[20] and to gossip about the recently deceased Madame de Staël and her writings. Rarely does one find so full a report of an evening in French scientific society as is Mrs. Somerville's. Particularly interesting is her account of Cuvier's comparison of French and Scottish schools:

. . . [M. Cuvier] said when he was sent by Government to examine the state of the schools at Bordeaux & Marsailles [sic] he found them in a wretched condition, few being able to read, and hardly any could perform a simple computation in arithmetic they acquired as much as was necessary to carry on commerce when they went into the counting house, as for Science history or literature they were unknown, and the names of the most celebrated of the French philosophers, famed in other countries, remain utterly unknown to those who dwell at a distance from Paris. Mr Biot having written home he found in Aberdeen not one, but many who perfectly understood the object of his voyage & were competent to converse with him on the subject. Mr Cuvier said that constituted one of the striking differences between the two countries, for it consisted with his knowledge that the like could not occur in any town in France. Science was highly cultivated but confined to the Capital. . . .[21]

On the following day — Sunday, 27 July — the Somervilles visited Lucas at his home to view his private collection of minerals.[22] He presented them with several specimens and a copy of his two-volume work on mineralogy.[23] The next two days were spent sightseeing with Scottish friends, but on Wednesday, 30 July, they viewed 'a splendid Collection of Minerals which had been the ruin of the proprietor (the Marquis de Drae) and . . . [was] now in the hands of his trustees for sale'.[24] The Somervilles themselves were interested in enlarging their collection. The visitors were 'next conducted thro' L'Ecol [sic] des Mines' by the mineralogist A. J. Brochant de Villiers (1773–1840), who showed them specimens of minerals from each department of France and '. . . a complete suite of rocks presented to L'Ecol by Werner'.[25]

That evening Madame Biot again arranged a dinner for them, this one 'on purpose to show . . . [the Somervilles] as she said les personnes distinguees'[26] — the Aragos, the Gay-Lussacs, the Thenards, and the Poissons (whose wedding had taken place the previous day). A number of persons, including Baron Alexander von Humboldt (1769–1859; F.R.S., 1815), came in after dinner and afterwards two card parties were set up — one a table of Boston, a form of whist that took its technical terms from the Siege of Boston in the American War, and the other of vingt-un.

The entry for the following day, 31 July, is the last in the bound journal. On that Thursday the Somervilles visited Vincennes, the Bastille, the cemetery of Père la Chaise, and a mirror factory.[27] There are in the Somerville Collection, however, a number of loose sheets covered with Mary Somerville's handwriting and clearly intended as notes to enable her to continue her diary.[28] They describe a combination of sightseeing, entertainment and social engagements. On 1 August, for example, the couple saw a display of farm machinery, watches at the shop of the famed horologist A. L. Breguet

(1747–1823), and 'Mme Curcier's mathematical library',[29] A 'Mme Ve Courcier' published in 1817–19 the third edition of Biot's translation of Fischer and may be this lady. The Somervilles dined that day with the Gay-Lussacs, the party including the 'Aragos, Humboldt, Tenard [sic] &c'.[30] The '&c' may have been 'M^r LaPlace and Mr Bertholet [sic]', whose names are jotted down just before mention of the dinner party.

On the following day — their eleventh in Paris — they dined with the Aragos, 'Mme Biot, Humbold [sic], Mr Gay-Lussac, Mr Mathieu &C.' also being present.[31] Madame Biot and Madame Arago had, shortly before the Somervilles' arrival in Paris, effected a reconciliation between their husbands[22] and so could accept each other's hospitality. Apparently the conversation touched on astronomy, for Mrs. Somerville's notes include the words, 'Great Telescope 187 feet . . . Repeating circles &c'. The evening was again spent with the Cuviers; among the other guests were Madame Biot, Humboldt, Prony, Poinsot, and Thomas Young. Young had visited a patient in Paris early in 1817 and been so 'pleased with his reception in the scientific circles of that metropolis [that] he went back . . . for a few weeks in the summer of the same year'.[33] Mrs. Somerville's only note about the evening reads, 'Conversations about Meridian'.[34]

Only one of the two visits the Somervilles made to Arcueil is described in these notes:

5th. Dined at Arcueil with M. LaPlace. party M & Mme Arago, M & Mme Bertholet and several Gent^n entertainment very handsome. M LaPlace said with regard to the Mech. Cel. it is probable that improvements may be made in analysis and that methods may be found to make the series converge whether the inclination of the planes of the orbits of the planets be great or small but at all events the work would remain a beau monument de l'esprit humain [sic]. I complimented him on the System du Monde as a work displaying at once depth of science and elegance of composition, he said it was very true but that he did not think the English or any other nation could appreciate the beauties of French literature — As a proof which Lady Morgan did not admire Racine. He told me he continues to work hard and that he had read a paper at the Institut the day before; [on = crossed out] paper [on the probability of error = crossed out] giving formulas to ascertain the probable amount of error &c in Geodasic [sic] new moment, or triangulation — Having mentioned that Dr [Thomas] Young had that morning read the Yrogliphic [sic] inscription on a mummy at the Kings library to us, he said he would believe Dr. Y. or any man knew the key if any one person agreed with him in the translation. The conversation having turned upon Egyptian monuments, M. Arago told that from the measures employed, in the Pyramid it was more likely that they had measured the Meridian. A wall 10 feet high & 1 foot thick. Adm des anglais tuer reconnaissable.[35]

Some 50 years later when writing her autobiography, Mrs. Somerville sometimes confused events laid out in the travel diary with those that took place on later visits, and she sometimes changed emphases. Her 1817 journal, for example, says little of Mesdames Arago, Cuvier and Laplace but refers frequently to Madame Biot; her printed autobiography, on the other hand, is exactly the opposite. Great names — Laplace, Arago, Biot, Gay-Lussac, Cuvier and the like are found in both, but minor figures, such as Lucas and Brochant de Villiers, disappear in the *Personal Recollections* while a few Parisians — the two medical men François Magendie (1783–1855) and J. D. Larrey (1766–1842) — unmentioned in the diary and notes are included in the autobiography.[36]

Her memoirs also dwell to a much greater extent than the journal on the personal appearances and characteristics of her Parisian hosts. Arago rather than Madame Biot emerges in the autobiography as their chief sponsor in 1817. He is described as

. . . tall and good looking with an animated countenance and black eyes. His character was noble, generous and singularly energetic; his manners lively and even gay. He was a man of extensive and general information and from his excitable temperament he entered ardently into the politics and passing events of the times as well as into science in which few had more extensive knowledge. On this account I thought his conversation more brilliant than that of any of the french savans with whom I was acquainted. . . .[37]

Madame Arago is described as 'gentle [and] ladylike'.[38]

According to the account in the autobiography, Arago introduced the Somervilles to Laplace. 'The Marquis was not tall but thin, upright & rather formal', Mrs. Somerville wrote, 'distinguished in his manners & [with] a little of the courtier in them perhaps from having been so much at the Court of the first Napoleon . . . [but] was exceedingly kind and attentive [to me]'. The Marquise was 'quite an elegante'. Mrs. Somerville found Laplace 'incomparably superior to Arago in mathematics and astronomical science . . . [but] inferior to him in general acquirements so that his conversation was less varied and popular.' Poisson 'had all the vivacity of a Frenchman'. Cuvier was 'under middle size, agreeable but conscious of the high place he held in the scientific world'.[39] The judgements Mary Somerville expresses are those of an experienced observer, moving easily in a society renowned for its achievements, and are much more consonant with the fame and renown of her last years than with the opinions likely to be voiced by a timid but fair-minded woman recently introduced into distinguished foreign scientific society.

The Somervilles could well be gratified by the success of their first Paris visit. During their short stay in the French capital they had become acquainted with important segments of the French scientific establishment and had seen many of its institutions and installations in operation. The couple left Paris on friendly footing not only with members of the influential Society of Arcueil — Laplace, Berthollet, Arago, Biot, Gay-Lussac, Humboldt, Poisson, Thenard (and likely Dulong also) — and their wives and families but with other notable savants, such as Cuvier and Haüy, and with many lesser figures. Mary Somerville, an alert onlooker rather than an active practitioner of science, had seized the opportunity to acquire scientific friends, information and experience that would stand her in good stead in coming years and that would enable her to speak with firsthand assurance of Parisian science, Parisian scientists and Paris itself. Moreover, this important scientific community now knew her personally.

2. Switzerland

'At Paris I equipped myself in proper dresses, and we proceeded by Fontainbleu to Geneva',[40] Mrs. Somerville wrote in her memoirs of the next stage of their Continental tour. This visit was her introduction to Switzerland, a country she came to love and one whose scenery she sought to capture over the years in sketch after sketch and painting after painting. Of all Swiss cities in these decades, Geneva had the most active intellectual life. Its men of science, with ties to Paris, Edinburgh and London, were among the most gifted on the Continent.

The Genevese savants — Mrs. Somerville lists only M. A. Pictet (1752–1825; F.R.S., 1791),[41] A. P. de Candolle (1778–1841; F.R.S., 1822), Gaspard de la Rive (1770–1834) and Pierre Prevost (1751–1834; F.R.S., 1806), along with the historian J.C.L.S. di Sisimondi (1773–1842)[42] as having been in Geneva at the time — received the couple cordially. These men were bound together by a shared interest in science and by ties of kinship and marriage. Prevost, for example, was the husband of one of Aleander Marcet's sisters. They and others were widely respected in the larger scientific world and had made Geneva — next to Paris — the strongest scientific centre in Europe.

The Somervilles, introduced by Alexander and Jane Marcet and fresh from a visit with Paris savants long associated with this group, were immediately welcomed by Genevese scientific society. They found there also another Scottish visitor, Professor John Playfair, abroad for his health. Playfair promptly resumed his old role of tutor to Mary Somerville. She wrote her mother:

> We had the good fortune to meet with Mr. Playfair and his nephew here. I thought we were quite at home when we saw him. he is in perfect helth. he and his nephew, Mr Pictet, William and I ascended a very high hill this morning which is very curious on account of its geology — We went at 7 in the morning had breakfast in the open air and returned between two and three — Tomorrow morning at 6 we set out for Chamonix and St Bernard the excursion will last eight days it is chiefly performed on mules so that it is impossible to take Marg^t. She remains at the Hotel with Mad^{lle} I have got shoes with nails, worsted stockings & everything ready for snow and ice, so in my next you will have a long account of the Glaciers. After this excursion to the Alps we begin our tour of Switzerland, for you know Mont Blanc is in Savoy. . . .[43]

Mrs. Somerville in her narrative of the week-long expedition to the Convent of St. Bernard[44] does not specify that Playfair was one of the party, but it is clear from her letters that she took every opportunity to enlarge her knowledge of geology and of mineralogy during her stay in Switzerland. Following their return to Geneva and a brief respite to recruit their strength, the Somervilles began what they considered an 'ordinary short tour of Switzerland'[45] — up the lakes to Bern and Interlaken. For Mary Somerville it was an opportunity to make numerous sketches of landscape and to observe at first hand geological, mineralogical and meteorological phenomena. Her notes and letters demonstrate her keen interest in such matters and give evidence of an acute eye, a retentive memory and a vivid pen; the clarity of her writing hints at talents for exposition that would be apparent later in her books.

In September, after two months abroad, the family started homeward. Whether this early return had been their original plan is unclear; William Somerville's Army status was still unsettled and there are vague references in family letters to a presentation of his 'case' in London.[46, 47] At any rate by the time they reached Vevey both Mary Somerville and her daughter were too ill to continue the journey. During the past weeks of constant — often excessive — activity and travel among strange surroundings, the health of the party had been remarkably good, but the fever which now struck was serious. Young Margaret recovered rapidly but her mother was barely convalescent two months later. From Lausanne an enfeebled Mary Somerville wrote her sister-in-law that only William Somerville's 'excellent medical skill under God . . . [has] saved both my dear Margaret and myself; I never saw any person have the talent so completely of nursing'.[48]

She was also indebted to some travelling English ladies, 'who . . . sat up alternate nights with me, as if I had been their sister'.[49] This meeting is characteristic of the good fortune that repeatedly attended Mary Somerville's career, for the ladies were the Misses Barclay, daughters of Robert Barclay of Bury Hill (1751–1830). A Quaker family related to the great Quaker families of England and Scotland, the Barclays of Bury Hill had several connections with science, one of the closest arising from the marriage of the fourth Barclay daughter in 1814 to the Cornish philosopher and inventor Robert Were Fox of Falmouth (1789–1877; F.R.S., 1848). From this chance encounter in Switzerland a lasting friendship developed. The Somervilles from time to time in the next years made visits to Bury Hill.

3. Italy

William Somerville, alarmed at his wife's state of health, decided that they should spent the winter in Italy, where her recovery could be completed in a warm climate and where he himself might find well-off and influential patients among the many English that flocked to Rome. Mary Somerville was delighted with his plan. Accompanied by the Misses Barclay — who were to pass some time in Milan — they crossed the Alps by the Simplon Pass and proceeded to that northern Italian city. From there in slow stages over the next month and a half, the Somervilles moved southward, stopping at

Bassano, Venice, Padua, Bologna and Florence. Letters of introduction insured that they met the best society in each of these cities and they were hospitably entertained, but rarely were their hosts scientific.

Unlike London, Edinburgh or Paris, unlike most of the larger provincial towns of Britain and the Continent, the various states and cities of the Italian peninsula offered little science at the time, and that done by one or two isolated scientists rather than in the specialist societies or institutions that characterized England and France. In 1817 Alessandro Volta (1745–1827; F.R.S., 1791), the Italian scientist best known in England, was living in Padua, but there is no evidence that the Somervilles met him as they passed through that city. Nor does Mary Somerville mention Domenico P. Morichini (1773–1836; F.R.S., 1827), whose laboratory Davy (1778–1829; F.R.S., 1803) and Faraday (1791–1867; F.R.S., 1824) had visited in 1814[50] and whose investigations of the magnetizing power of the violet rays of the sun continued to arouse sufficient interest in Britain to be reported in the *Scots Magazine* of August 1817.[51] Some years later she and Morichini would be in correspondence[52] and almost a decade later she herself would undertake a similar experiment.

Although the Somervilles during their months in Italy met scores of people — many of them highly distinguished — only two of the number, on present evidence, can be considered scientific: Signore Nobili Alberto Parolini of Bassano and Professore Giovanni Battista Amici (1786–1868) of Modena. They became acquainted with Parolini, a keen naturalist and horticulturist, while in Paris, and he insisted they visit his northern Italian estate should they be in its vicinity. At Bassano he maintained a 'very pretty botanical garden'[53] and was their attentive and delightful host during the three days they were there. Mrs. Somerville says nothing of any geological interests on his part at the time, but after William Somerville's return to London, Parolini was made a Foreign Member of the readily welcoming Geological Society of London.[54] One of his daughters later married the English Alpinist John Ball (1818–1889; F.R.S., 1868).[55]

It was on their journey back to England that the Somervilles in June 1818 spent a few days in Modena, where Amici supplied them with a camera lucida like the one with which he made their likenesses and showed them his 'excellent instruments', including his powerful new catatropic microscope which rendered 'visible [to them] the circulation of the sap in the very minute portion of aquatic vegetable [Chara]'.[56] For more than a decade this gifted scientist, trained as an engineer-architect at Bologna, had taught mathematics at the Modena *liceo* while pursuing his chief interest, the manufacture and improvement of optical instruments. His new microscope, superior to any yet available in England, would that very year make him famous as an optician and as a microscopic botanist. The Somervilles also met his wife, who read and spoke English, and a lasting friendship between the two couples began.

In all likelihood the returning travellers reached Geneva shortly afterward and there saw Alexander and Jane Marcet, an encounter to which Mary Somerville makes perplexing reference in her autobiography. More than

50 years after the event she wrote that in this Swiss city on their way *to* Italy in 1817, she and William Somerville '. . . found Dr. Marcet, with whom my husband had already been acquainted in London [and] I, for the first time, met Mrs. Marcet. . .'.[57] She misremembers on two points: their first meeting had already taken place in London in 1816 and the Marcets were not in Geneva in 1817. Marcet returned to his native home in mid-June 1818 for the first time since his exile 24 years earlier and brought his wife with him to visit family and friends he had left behind.[58] Doubtless the Somervilles and the Marcets did meet in Switzerland in 1818 and doubtless also Mrs. Somerville is correct when she states that '. . . through these kind friends we became acquainted with . . . De Candolle, Prevost, and Da la Rive',[59] but introductions to these Genevese savants made by the Marcets in 1817 must have been by letter.

The Somervilles' letters from Italy in 1817–18 mention natural phenomena observed — volcanic eruptions, meteorological oddities, new plants and animals — but say little of technological processes or experimental science. They are filled with descriptions of music, opera, antiquities, paintings and sculptures, along with accounts of the exciting scenes, sounds and smells of Italy, its vibrant people and customs, and its magnificent scenery and architecture. Warmly received by the best society wherever they stopped, the couple glimpsed Lord Byron at one of the famed Venetian literary receptions of the Countess Albrizzi,[60] whom they had recently encountered in Paris. In Bologna they spent several hours with the 'walking polyglot', Professore Guiseppe Mezzofanti (1774–1849),[61] and in Florence they met the supercilious and insolent Countess of Albany (1742–1824),[62] widow of Charles Edward Stuart.

In Rome, which they reached before Christmas and where Dr. Somerville soon established a numerous practice among the English colony, the Somervilles made several friends who would figure in their future lives. Among them was young Captain Roderick Impey Murchison and his wife Charlotte. Despairing of further advancement after Waterloo, Murchison, newly married, had retired from the army and was now passing his time in travel and other light pursuits. Of him Mary Somerville wrote:

. . . At that time he hardly knew one stone from another. He had been a cavalry officer in the Peninsular War, was an excellent horseman and like other young men of his rank and age a keen fox hunter. Mrs. Murchison was an amiable accomplished woman, drew prettily & what was rare at the time she had studied science, especially geology and it was chiefly owing to her example that her husband turned his mind to those pursuits in which he afterwards obtained such distinction. I then formed a friendship with her and her husband which neither time nor distance has interrupted.[63]

In addition to sightseeing and social affairs, to hearing 'good music' and being presented to Pope Pius VII (1742–1823), Mary Somerville busied herself while in Rome with learning Italian, visiting galleries,[64] and with writing down the accounts given by the antiquary whom they engaged as guide.[65] There and in Naples, where they spent their last two months in Italy, visits to antiquities and natural wonders were her chief amusement. The Somervilles toured Pompeii and, with young Margaret, climbed to the top of Vesuvius, which had erupted just before their arrival.[66] In Naples they made another fortunate friendship, this one with the eccentric and wealthy Miss Patty Smith,[67] spinster daughter of the Norwich liberal and M.P. William Smith (1756–1835; F.R.S., 1806). Mrs. Somerville always ascribed the quick ripening of her friendship with the Smith family to the 'bond of interest' created by their common stand against slavery. She and Patty Smith also shared other concerns. Miss Smith was 'liberal in her opinions, witty original, an excellent horsewoman . . . drew cleverly . . . was a good judge of art' who taught Mary Somerville much '. . . with regard to pictures and style of drawing' among the old masters,[68] as well as introducing her, on their return to London, into circles — particularly the Clapham House Sect — with whom the Smiths were intimate. Mrs. Somerville knew Patty Smith's niece, Florence Nightingale, as a little child and regretted that she did not meet her again in later years.[69]

By mid-April 1818 Mary Somerville was beginning to long for home. They had been abroad for almost ten months. She wrote her mother '. . . I rejoice in the prospect of our return . . . and am delighted to turn North perfectly satisfied with what we have seen'.[70] The months abroad had been a time of glorious development for this talented woman; their harvest would come in later years.

4. The Return

Although details are missing — there are no letters or papers in the Somerville Collection dated between 12 April 1818 and 23 February 1819 — the Somervilles appear to have returned to England by the late summer or early autumn of 1818, gone to Scotland for a family visit,[71] then settled again in London with their young daughters. It was apparently to London that Madame Gay-Lussac[72] sent information about a possible nursery maid for the little Somerville girls in February 1819. Another letter, dated 5 June 1819 and written in London, offers lengthy and detailed advice to Mrs. Somerville on her botanical studies. Her correspondent is the Genevese botanist de Candolle,[73] in England on a visit.

But it is a third letter — this one from William Somerville to G. B. Amici in Modena[74] — that firmly fixes the Somervilles at their favorite London residence, No. 12, Hanover Square, by mid-year and indicates that they had resumed their intercourse with scientific society. In it Somerville recalls their stay in Modena and offers a warm welcome whenever Amici comes to London. Desirous '. . . that so useful an Instrument should . . . find its way to England', he orders one of Amici's new microscopes and requests guidance as to its use, then mentions that Kater has lately reported to the Royal Society (to which Somerville had been elected a Fellow on 11 December 1817) important work on ascertaining the figure of the earth with his pendulum. Clearly the Somervilles intended to maintain and utilize the ties established by them on their Continental tour.

Amici replied promptly, his cordial letter[75] divulging some of the history of early nineteenth-century microscopes in Britain. He pointed out that he had no order from London for his instruments other than one from the Austrian ambassador, Prince Paul Esterházy (1786–1866), and that he could easily send one to the Somervilles in April 1820 along with Esterházy's. Further he asked Dr. Somerville to deliver in person to Sir Joseph Banks the printed description of the new microscope and the accounts of its use which the inventor had prepared. The Somervilles thus became owners of one of the first two Amici microscopes in England.

The Hanover Square establishment to which it was sent was much closer to the centres of organized scientific life in the English capital than had been their house in Queen Square. Although expensive, the new residence was otherwise well suited to their needs. A letter from Thomas Somerville,[76] dated 17 July 1819 and addressed to No. 12, Hanover Square suggests, through its extolling the advantages of good conduct, industry, frugality, and self-denial, that the matter of William Somerville's employment was still unsettled at that period. Well might an aged father be concerned, for his physician son was now in his forty-ninth year, with all the tastes of a gentleman yet without fortune or prospects of inheritance, with no secure post, and with a wife and three small daughters to support and a grown son ready to be launched into the world. When on 28 December 1815 William Somerville had been gazetted a Principal Inspector of Hospitals in the Army Medical Service, his future had seemed bright. By Christmas of the following year, his new post had been abolished, he had been lucky to gain half-pay,[77] and his future seemed bleak. Neither in 1817 nor in 1818 did his name appear on the Army Lists. By the time the family was back in London in 1819, he had accepted the fact that he must establish a private medical practice or else look about for another appointment.[78] In the latter venture he was unexpectedly successful.

When around mid-September he learned that the Physician to the Royal Hospital at Chelsea, Dr. Benjamin Moseley (1742–1819), was dying, Somerville instantly set about winning the soon-to-be-vacated post. He turned first to an old friend at the Horse Guards, Major-General Sir Henry Torrens (1779–1828), for a recommendation to the head of the army, the Duke of York. Torrens was not in town when Somerville heard of the impending vacancy, but the doctor — travelling more than a hundred miles by coach in less than 24 hours — finally located him in the country and readily obtained the desired letter from this chief dispenser of army patronage. When he delivered it in person to the Duke of York, however, Somerville learned that the appointment was civil rather than military. The Duke nonetheless promptly wrote, on the doctor's behalf, to the paymaster general who, in turn, quickly recommended that Somerville be appointed 'immediately upon the decease of Moseley'.[79]

Thus the patronage for which William and Mary Somerville had so long hoped at last came their way. Dr. Somerville had succeeded in obtaining what his wife described as 'the one appointment in London to which he is elegeable [*sic*]'.[80] Benjamin Moseley died on 25 September. Nearly two months later, on 13 November 1819, William Somerville was officially gazetted physician to Chelsea Hospital.[81] It meant, as Mary Somerville wrote her father-in-law, a 'salary [of] about £300 per ann., a good house, and a garden with coals, candles and no taxes . . . the [total] amount we estimate . . . at about £700 or £800 a year. Better still it in no way interferes with practice but rather will promote it . . .'.[82]

This office, a semi-sinecure which William Somerville held for almost 21 years, was the apex of his medical career. He seems to have been competent and conscientious in filling it, but his medical light shines dim beside that of such talented contemporaries as Charles Bell, Richard Bright, Marshall Hall or Thomas Addison. Nor did he earn the generous fees that characterized the practice of fashionable London doctors such as Astley Cooper, W. F. Chambers, C. M. Clarke, Henry Holland or even the exotic A. B. Granville. Somerville's gift was for nursing. He had a store of common sense and, by 1819, a fairly wide medical experience but no medical genius. He showed no interest in collecting or publishing his numerous medical observations.

There is no evidence that William Somerville, who straightway joined several scientific societies on coming to London, was ever a member of any of the London specialist medical societies such as the Medical and Chirurgical, of which his friend Alexander Marcet had been a founder. Somerville appears to have occupied no more than a peripheral place in the London medical establishment, to have consistently been at several removes from important medical work and medical discoveries in his own time, and to have been undisturbed by his position in the medical world. He enjoyed escorting medical men around his hospital and impressed at least one distinguished military doctor with his store of practical knowledge and inventiveness.[83] To visitors in London he was unfailingly helpful, foreigners especially appreciating his useful flair for languages. His election in March 1834 to the Académie Royale de Médecine of Paris[84] shortly after his wife's lengthy and triumphant stay in the French capital and after a visit to Chelsea by the famed French medical authority Antoine Clot ('Clot Bey', 1795–1868) may have been at least partially in return for Somerville's kindnesses to visitors as well as an indirect tribute to his wife's fame. No equivalent British medical honour ever came his way.

William Somerville's warmest enthusiasm was reserved for his avocations. He seems to have regarded his profession — useful as it was in advancing his own standing in society and in furthering his wife's scientific pursuits —simply as a means of livelihood, one that was often inconvenient and a bother but a necessary chore. The convivial aspects of the profession, however, appealed to him. He was a member of at least one medical dining club — the exclusive Pow-Wow Club, founded by John Hunter and numbering among its members Sir Gilbert Blane, Sir James McGrigor, Henry Holland, Sir Astley Cooper, John Yelloly, Sir Walter Farquhar, P. M. Roget and Sir Everard Home; it dissolved in 1830.[85] He also belonged to one of London's several Beefsteak Clubs and dined often in that group.[86]

William Somerville found more in common with men like Wollaston and Thomas Young — both of whom had given up medicine for science — or with men whose chief interest had always been scientific than he found with medical people, apart from those in the military. In the army medical establishment, station and ability to get on with officialdom were more important in gaining advancement than was medical skill. As an army doctor William Somerville remained on good terms with his superiors, benefited from the good will Farquhar, Blane and McGrigor always displayed toward him, and gained position, income — though never enough for his needs and wishes — and opportunity to rise in the world. Flexible working arrangements and an easy policy of official leaves made long absences from his post possible. Even so, he and his family complained at times of the burden of official duties.

There is never a hint in the Somerville papers that he — a man eager for place and income — ever saw his own medical skills as a realistic means of achieving these goals. Rather, he used official position, friendships and personal qualities to advance his own standing and to further his wife's scientific pursuits. William Somerville appears to have carried out his professional duties adequately and to have hoped that through some stroke of good fortune he would become a gentleman of leisure and influence.

By 1819 the Royal Hospital — or Chelsea College as it was often called — housed some 500 pensioners, all invalided or retired soldiers. Staff houses and other structures designed by John Soane had recently been added to the Christopher Wren buildings within the spacious grounds of the Hospital. The head of the establishment was a governor — always a retired officer of high rank rewarded with this post. Second in command was a lieutenant-governor in the same mould. Two younger military men served as adjutant and major, respectively. A chaplain, a physician, and a surgeon completed the official roster. At the time of Somerville's appointment, Thomas Keate (1745–1821), a respected medical man, was Surgeon. After his death the post was given to the officious Sir Everard Home (1756–1832; F.R.S., 1787).

By the end of 1819 Somerville had taken up his new appointment and the Somervilles were ready to begin a new London life.

CHAPTER 4

IN THE MAINSTREAM OF LONDON SCIENCE

1. Scientific Training in the 1820s

Mary Somerville's earliest scientific intimates in London were William Hyde Wollaston (1766–1828; F.R.S., 1793); Thomas Young (1773–1829; F.R.S., 1794); Henry Kater (1777–1835; F.R.S., 1814); Sir John Saunders Sebright (1767–1846); Francis Chantrey (1781–1841; F.R.S., 1818), William Blake (1774–1852; F.R.S., 1807) and Henry Warburton (1784–1848; F.R.S., 1809). Alexander and Jane Marcet were part of this circle until Marcet's death in October 1822 and his wife's removal to Geneva. The Somervilles were the newcomers to this group, which had drawn together with Wollaston as the magnet. He, Young, Blake and Warburton had all been at Cambridge. After the latter two settled in London, they turned to science as an avocation. All were members of the Royal Society, Wollaston, Warburton and Blake of the Geological Society. Wollaston and Young often served on committees of the Royal Society and as officers. Sebright, according to Mrs. Somerville, was 'a good chemist'.[1] Chantrey was interested in mineralogy.[2] He had become a Fellow of the Geological Society in 1814. All were keen sportsmen. Kater became part of the group in 1814 when he came to London and joined the Royal Society. Wollaston and Young immediately recognized his talents and made sure he took an active part in the management of the Society.

The Somervilles found a community of interests with this group. Jane Marcet and Mary Frances Kater — who assisted her husband with his calculations — both had families of young children. Neither had undertaken a serious study of science as Mrs. Somerville already had done of mathematics. These two ladies may have accustomed the gentlemen of this little group to scientific women but neither showed the talent or commitment that Mrs. Somerville did. The willingness of these philosophers — especially the three who devoted all their time to science, Wollaston, Young and Kater — to accept Mrs. Somerville as their pupil lies in great part in her own serious approach to her studies and in the flair she displayed for them. That such informal training as they gave their protégée could by 1825 produce a practicing scientist arose from the then current state of science and science education. Her personal cultivation of science mirrors a more general cultivation of a study that, as the century progressed, attracted more and more notice, won more and more adherents.

The Scientific Establishment which the Somervilles found on their arrival in London in 1816 and into which they so readily fitted was dominated, as it had been for well over a century, by the venerable Royal Society of London, comprehending all the sciences. Four much younger and less influential

bodies directed attention to certain sub-branches of science. Three of them
— the Linnean Society (founded 1785), the Horticultural (1804), and the
Geological (1807) — were flourishing but a fourth — the Society for Promoting Natural History (1782) — was in decline and in 1822 would dissolve. The
Royal Institution, established in 1799 to encourage practical application of
science, soon became part of the Scientific Establishment. Its public lectures
drew hundreds of fashionable men and women caught up in the passion for
science to Albemarle Street, while brilliant discoveries in its laboratories
added lustre to British science. These lectures would be a source of information and inspiration to Mary Somerville.[3]

Membership in these various groups tended to overlap extensively. Meeting rooms were often shared and meeting dates chosen to accommodate
multiple fellowships. Medical men of the capital traditionally took prominent
roles in the Royal Society and later in the specialist societies (including the
specialist medical societies that arose from 1805). Actual management of each
organization rested in the hands of a small group of dedicated members; these
councils were in effect self-perpetuating bodies.

The companionable aspect of the London societies is notable. Dining clubs
were often associated with specific societies; the clubs tended to be small and
selective and met regularly at a chosen London tavern. Guests were frequent;[4] a visiting philosopher could expect one or two invitations during a
stay in London. Science in the 1820s was a sociable enterprise. Discoveries
might be made in isolation but their reporting took place in a club-like
atmosphere. A man's willingness to entertain his fellows, for example, was an
important factor in his selection as president of the Royal Society.[5] William
Somerville, gracious and affable, not only enjoyed this aspect of London
science but added to it.

Few fellows were practising scientists. A larger number were friends or
patrons of science. The majority, however, joined a society through impulse
or ambition or amiability. Indiscriminate admission to membership would
become an important issue in 1830. Already winds of change were stirring.
The earliest breeze was felt when the Geological Society asserted its independence soon after its founding and civilly refused to become, as Sir Joseph
Banks (1743–1820; F.R.S., 1766; P.R.S., 1778–1820) wished it to, an inferior
subsidiary of the Royal Society. Next came establishment of a separate
astronomical society in 1820. Both groups attracted strong memberships,
directed attention to their specialities and served as important supplements,
not rivals, to the Royal Society. The great burst of scientific activity from the
beginning of the century led, especially from 1820, to the establishment of
other specialist societies in the capital and to a good number of provincial
bodies. Mary Somerville found her principal scientific associates among
men who had taken an early part in this development in organized science.
Venturesome enough to strike out in these new directions, they were venturesome enough also to applaud a woman's interest in science.

The philosophers who pressed for the foundation of new specialist societies
were for the most part men who were gradually reassessing their own roles

and the role of science itself. Not yet ready to insist on full-fledged professionalization of science, they were moving, through their own efforts to reform the management and functioning of the Royal Society and to modernize the national perception of scientific undertakings, in this direction. The studies which they had begun as avocations had come, for many of them, to occupy their full attention. The standards which they imposed upon themselves and looked for in others were far above a mere dilettante interest. Several philosophers devoted all their time to scientific pursuits but did not stand apart from their fellows who were merely friends of science. Instead they shared their enthusiasms and their researches. No adjustment was required in order to accept Mrs. Somerville as one of their eager listeners. These practitioners demanded no rigid qualifications or course of preparatory study for a rising philosopher — indeed, there were none. They did insist on a true commitment to science, one without affectation or frivolity.

Scientific training was still largely an apprenticeship entered into for love of the subject. A few scientists gained a livelihood through teaching or other scholarly professions, a few military and naval men carried out scientific investigations as part of their duties, the medical men had their patients, and a few scattered practitioners found posts at the Royal Institution or the British Museum or similar establishments, but no youth entered science as he might enter the church or law or medicine to support himself and a family. A few philosophers — Wollaston most notably — gained fortune through their discoveries, but opportunities for sudden wealth in the 1820s lay mainly in overseas adventures and in the booming domestic industrialization and rapid technological change.

In the informal apprenticeship that produced a coming scientific practitioner, a master philosopher through advice and example guided the novice into full participation in his speciality. Discussion of scientific principles and findings, observation of scientific activities, and criticism of scientific efforts were the chief tools of instruction. The experienced practitioner directed the learner's readings, showed him how to use apparatus and how to design experiments and instruments, and introduced him to the scientific community. Through mutual exchange scientific knowledge was spread and advanced.

The common goal of master and pupil was to add to the sum of human knowledge. There might be other rewards — fame, position, wealth — and there was the satisfaction of doing a job one valued, but these were held to be incidental, merely the deserved recognition and reward of efforts benefiting the nation and mankind. Natural philosophers felt a deep obligation to instruct those who earnestly sought their help.

Mary Somerville thus came to her training in science in much the same way as any male in these years would enter these studies. In the nurturing atmosphere that obtained in London, latent talents such as hers could be cultivated and could flourish. It was her good fortune to find herself, as soon as she reached London, among men who were able and willing to guide her, ready to encourage the use of her obvious and considerable natural gifts, and to pay homage to her accomplishments. It was her further good fortune to find others through the coming years who would carry on her development, so that for well over half a century she was closely associated with many of the best minds in England and abroad.

It was through simple social introductions that Mary Somerville met these eminent men of English science in 1816 and through social gatherings that her intimacy with them was established. For the greater part of her scientific life, this pattern would prevail. Serious philosophers quickly came to perceive her not only as a charming companion but as an apt and serious pupil.

The pace of life in the Somervilles' circle in the 1820s was leisurely. There was time for morning calls, for small evening parties, for breakfasts and dinners, concerts, the opera, and the theatre. Occasionally there were grand balls and public spectacles. The gentlemen had their dining clubs and their learned societies, and — for a few —their experimenting and their writing. For them, from time to time, there might be official responsibilities on committees or as office holders. A few served in parliament, others had businesses to oversee. Family affairs and management of country places required attention. The ladies had their domestic and maternal responsibilities but for the greater part of the day were free to enjoy themselves in society or to pursue whatever studies or amusements they wished. For both men and women there was time for holidays and travels and for leisurely house-parties in the country. Above all, there was time for science, science that could be shared and savoured, that carried not only the aura of discovery but that of enlightened entertainment.

In such circles and under such circumstances, and with the best tutelage which was available, Mary Somerville from 1820 to 1825 learned the fundamental principles and methods of physical science. By 1825 she was ready to apply them under the guidance of several foremost scientists of the day.

2. Mary Somerville's Apprenticeship

Mary Somerville's scientific instruction by these philosophers took place informally and in the midst of other activities. One passage in her autobiography suggests the way in which she was taught:

> Somerville and I used frequently to spend the evening with Captain and Mrs. Kater. Dr. Wollaston, Dr. Young, and others were generally of the party; sometimes we had music, for Captain & Mrs. Kater sang very prettily. All kinds of scientific subjects were discussed, experiments tried and astronomical observations made in a little garden in front of the house. One evening we had been trying the power of a telescope in separating double stars till about two in the morning; on our way home we saw a light in Dr. Young's window, and when Somerville rang the bell, down came the doctor himself . . . , and said, "Come in; I have something curious to show you . . . [about a hieroglyphic interpretation he had just made].[6]

She says nothing of instruction from Young but does give an account of the general difficulties he had encountered in gaining acceptance of his undulatory theory. Although her papers give no clue as to how her allegiance to this view came about, she was adamant in defence of Young's theory.

A letter from Kater, written in 1821 and accompanying his gift to her of a pocket polariscope, outlines 'an easy way of approximating the meridian', a method he has taken, as he points out, from the respected Brinkley's *Astronomy*.[7] At this time Mrs. Somerville appears to have been learning to use simple astronomical instruments; Kater refers to her transit.[8]

Wollaston was her earliest and chief London mentor. In her autobiography Mary Somerville gives an account that not only expresses her own opinion of him but illustrates the way in which she learned science. She and her husband had bought a good many mineral specimens while abroad. On their return they spent pleasant hours arranging them. Dr. Somerville used a blowpipe in analyzing the minerals; Mary Somerville did not learn the technique. Instead she made small cups to hold the specimens. Their first close acquaintance with Wollaston came about through these minerals and a 'lasting friendship' resulted. She wrote:

. . . [Dr. Wollaston] was gentlemanly, a cheerful companion, and a philosopher he was also of agreeable appearance, having a remarkably fine intellectual [looking = crossed out] head. He was essentially a chemist (and discovered Paladium [*sic*]) but there were few branches of science with which he was not more or less acquainted . . . His characteristic was extreme accuracy which particularly fitted him for giving that precision to the science of crystalography [*sic*] which it had not hitherto attained; by the invention of the goniometer, which bears his name, he was enabled to measure the angle formed by the faces of a crystal by means of the reflected image of bright objects seen in them. We bought a goniometer & Dr. Wollaston who often dined with us taught Somerville and me how to use it by measuring the angles of many crystals during the evening I learnt a great deal on a variety of subjects besides crystalography [*sic*] Dr. Wollaston at his death left to me a collection of models of the forms of all the natural crystals then known.[9]

One of the seven notes and letter from Wollaston in the Somerville Collection gives an example of his meticulous instruction. It explains to Mrs. Somerville in great detail the geometrical construction that will convert a plane surface, rolled into a cylinder, into a 'true panoramic view'.[10] Panoramas — painted pictures made to unroll continuously before a spectator — had been the rage in London for some years. Wollaston had a gift of taking some experience immediately at hand and presenting it in scientific terms. Mrs. Somerville tells how on one journey with them, he exclaimed as they passed some trees, 'Look at that ash tree; did you ever notice that the branches . . . are curves of double curvature?'[11]

In addition to his 'conversation on a variety of subjects, scientific and general',[12] he sent her books and papers[13] and lent her a variety of small scientific instruments,[14] teaching her to use them. He gave her, for example, a little prism made by Fraunhofer;[15] using it, Wollaston had independently discovered the dark lines across the solar spectrum. In his letters offering scientific advice and information there is no touch of condescension; he writes in a matter-of-fact way suitable to an exchange between scientific colleagues. From her contacts with him and with other men of learning, Mary Somerville gained authority as well as education. Later she could report their findings and opinions with the assurance that comes from having heard at first hand and being able to ask questions and make certain understanding was complete and correct. Wollaston to her was not only a valued friend and teacher but the supreme authority in the 1820s on science.

William Somerville's interest in science, especially mineralogy, and his growing intimacy with this particular circle — one powerful in the management of the Geological Society — gained for him in 1820–21 a place on the Council of that society.[16] He served only one year and had no other official post in science, save in the Royal Institution hierarchy, for over a decade. Nor did he ever achieve sufficient distinction or place to be made a member of the Royal Society Club, with whom — as the guest of Dr. Wollaston — he dined for the first time in 1821.[17] Club records show that he was again a guest in 1824.[18] Although he was on excellent terms with many scientists and was often an effective intermediary in scientific exchanges, always ready to interest himself in the internal affairs of his various associations, he seems to have preferred — or been relegated to — an unofficial role in most cases.

Beechwood Park, the Hertfordshire seat of Sir John Saunders Sebright, was often the scene of battues for this scientific band and for other distinguished guests. The Somervilles first went to Beechwood in late 1820.[19] Both Wollaston and Davy were guests on this occasion. Although the election in the Royal Society in November 1820 — which involved both Wollaston and Davy — had earlier agitated many of Wollaston's friends — among them the Katers and two younger philosophers, John F. W. Herschel (1792–1871; F.R.S., 1813) and Charles Babbage (1792–1871; F.R.S., 1816),[20] who were a junior contingent on the periphery of the Somerville circle — Mrs. Somerville says nothing of the affair. An issue in this election would re-appear in 1827 and again in 1830 and have a profound effect on English science in the first half of the century. The major question was whether English science could best be promoted by putting at its head a working scientist, familiar through experience with the needs, responsibilities and potential of modern science, or instead a distinguished patron and friend of science who could through his own station and influence act on its behalf. After the death of Sir Joseph Banks on 19 June 1820, the Society faced this issue for the first time in four decades. When the decision favoured a working scientist, the choice was between Wollaston and Davy. Wollaston had strong support, Davy was keen for the office. In the fortnight following Sir Joseph's death, discussions, meetings and manoeuvering reached fever pitch. Wollaston, who disliked contention, stepped aside early in July. He was, however, made interim president until the regular election in November, when Davy was installed.

The Somervilles knew Davy and saw him on occasion.[21] Mrs. Somerville admired his scientific brilliance but, like others of her intimate circle,[22] considered him at times ridiculous in his posturings and self importance. Lady Davy was an old Edinburgh friend.

The novelist Maria Edgeworth (1767–1849) was at Beechwood in January 1822 with a party that included Wollaston, the Somervilles and Jane Marcet. In her letters written at the time she gives a vivid account of this group.[23] Of the Somervilles she wrote:

Mrs. Somerville — the famous learned Mrs. Somerville — nothing learned or famous in her appearance — little slight-made — about the size of Mrs. Barbauld[24] and a flattering likeness of what Mrs. Barbauld at her best and most youthful days might have been — only she has no *set* smile or prim look — no *mimps* with her mouth — fair and fair hair with pinkish Scotch color in her cheeks — eyes grey — small round intelligent smiling eyes uncommonly close together — Very pleasing countenance — remarkably soft voice though speaking with a strong Scotch pronunciation — yet it is a well bred Scotch not like the Baillies. She was dressed in geranium colored Chinese crape. She is timid not disqualifyingly timid but naturally modest with a degree of self-possession through it which prevents her appearing in the least awkward and gives her all the advantage of her understanding at the same time that it adds a prepossessing charm to her manner and takes off all dread of her superior scientific learning. In talking to her we forget that La Place said she was the only woman in England who could understand and who could *correct* his works. She puts me in mind of Harriet Beaufort[25] more than of any one I ever saw. In the course of last night and during a walk this morning I have become perfectly at ease with her. Indeed our mutual friends Mrs. Marcet Dr. Holland and Mrs. Baillies[26] had prepared the way for our liking each other. Mrs. Somerville and Mrs. Marcet have undertaken to find me a first floor in some house half way between their houses in Harley St. and Hanover Square and they have invited us to come to them at all hours begging that we may drop in morning and evening so as to rest and make a home of their homes. Now as Dr. Wollaston and Dr. Holland and Mr. Warburton and Kater and all the best scientific and literary society in London drop into these two houses daily this privilege I consider most valuable. Dr. Somerville though not equal to his wife yet is in my opinion more agreeable than Dr. Marcet though not so pretty a man. He is round red Scotch faced and Scotch accented — has been in America — Italy — France and knows people and things well.[27]

From the early years of their marriage Mary Somerville was the magnet, William Somerville her willing satellite. Two decades later Maria Edgeworth, again in London, wrote her sister Harriet,[28] who had also been one of the Beechwood party, that Sydney Smith, discussing literature and science with her, had remarked:

. . . Mrs. Somerville — Her Introduction[29] to all those things about Light and stars and nebulae of which I [Smith] know nought . . . but her style clear and excellent — But as to herself I never could get anything out of her beyond what you might get from any semptress. She avoids all depth in converse — and no pretty superficial — very amiable no doubt. I'd bear the husband for her — that all I can say. Where is she?[30]

Science was a strong common interest among all the Edgeworth family, who moved happily in scientific circles in Ireland, England, France and Scotland. Maria Edgeworth and Mary Somerville became devoted friends and were much together during Miss Edgeworth's 1822 stay in London. She, who knew scientific society well, thought theirs 'the most agreeable as well as [the most] scientific' in London.[31] The Edgeworths considered Mrs. Somerville a better scientific hostess than the ones they had met in France, including Madame Cuvier.[32]

In the summer of 1822 the Somervilles and Wollaston journeyed together to Scotland. Edinburgh was agog with excitement at a visit of the new monarch, and they were in the Scottish capital for the festivities surrounding George IV's triumphal stay. Davy and the Marcets were also among those who made their way northward. Edinburgh society, especially scientific circles, was warmly hospitable.[33] Before taking a short tour of the Highlands,[34] the Somervilles and Wollaston visited Thomas Somerville at Jedburgh.[35] It was on this occasion that a Roxburghshire neighbour, their old friend Sir Walter Scott (1771–1832), arranged the famous angling party for Wollaston and Davy.[36]

Maria Edgeworth's letters and those of Alexander Marcet, together with the Somerville papers and Mary Somerville's autobiography, reveal how constantly the Somervilles' intimate London circle were in each other's company in the 1820s — at house parties in Hertfordshire (where Blake and Sebright had country estates) and on excursions, at London concerts and London events, at breakfasts and dinners and evening parties. Mrs. Somerville became famous for small evening gatherings at Hanover Square,[37] parties which she described as 'very entertaining' because she and her husband were 'intimately acquainted with every person of note both Scientific and Literary in London'.[38]

Close and frequent contact, especially the innumerable daily calls, was possible among this circle because all had London residences within easy walking distance of each other, of the learned societies and clubs frequented by the gentlemen, and of the Royal Institution in Albemarle Street, where Mary Somerville herself regularly attended the lectures.[39] They also had the leisure and the means to maintain such an existence. Wollaston, Young, Warburton, Blake, Chantrey, Sebright and Marcet were men of wealth, their fortunes earned through their own exertions and talents or acquired through inheritance or marriage. Kater and Somerville, although without fortunes, had secure — albeit relatively modest — incomes from their professions.

The everyday lives of this little band were filled with the social activities generally obtaining in London intellectual circles. Over the years their joint social and scientific lives became so interwoven as to create a single fabric of mutual concerns, preferences and entertainments. This close group was in turn part of a larger London circle that encompassed literary, artistic and political figures as well as philosophers.

Only hints of the relations between the Somervilles' early scientific circle and the engineering and technological giants of the day emerge from the Somerville papers. A number of famous engineers — Marc Isambard Brunel (1769–1849; F.R.S., 1814) and his son Isambard Kingdom (1806–1859; F.R.S., 1830), John Rennie (1761–1821; F.R.S., 1798) and his sons George

(1791–1866; F.R.S., 1822) and John (1794–1874; F.R.S., 1823) and Thomas Telford (1757–1834; F.R.S., 1827) — were Fellows of the Royal Society in the 1820s and 1830s and at its meetings were in the company of the gentlemen of the Somerville circle. Yet there are no letters from them in the Somerville Collection, and Mary Somerville mentions only two of them — I. K. Brunel and John (later Sir John) Rennie — in her autobiography. She recounts how, at Brunel's invitation, she and her husband twice made the much envied excursion beneath the Thames 'to see . . . [the] progress' of the excavations for Brunel's tunnel.[40] These descents must have taken place in the years 1825 to 1827. In 1830 William Somerville's name was the third one affixed to Brunel's Royal Society certificate. Among its fourteen other signers were nine close Somerville associates: Kater, Warburton, Chantrey, A. L. Wollaston, Faraday, Buckland, Fitton, Broderip and Basil Hall,[41] suggesting that Brunel was on familiar terms with the group.

When the second pier of the new London Bridge was under construction in the 1820s, its builder John Rennie gave the Somervilles 'an order of admission' so that Mary Somerville could 'see a coffer dam or frame in which the piers . . . [were] built'. In the manuscript of her autobiography she details her observations on this occasion.[42] Both manuscript drafts of this work have a lengthy and admiring passage about engineering wonders and name several engineers and manufacturers[43, 44] but this segment was omitted in the printed version. She particularly praises George Stephenson (1781–1848) for his character and perseverance and for his invention of railways[45] and of a safety lamp.[46] 'No man has had such influence or produced so great a revolution in the civilized world' as he, she declared.[47] His son Robert Stephenson (1803–1859; F.R.S., 1849) is hailed for his magnificent suspension bridges at Newcastle and over the Menai Strait.[48]

Locomotives, railways, steam vessels, tunnels, bridges, power machines and the telegraph are all mentioned in these manuscript pages, as are the names of several important tool engineers: Joseph Bramah (1748–1814), Henry Maudsley (1771–1831) and James Nasmyth (1808–1890). Mrs. Somerville omits to identify the last as the son of her Edinburgh painting teacher, the landscapist Alexander Nasmyth (1758–1840). In no instance in her autobiography or papers is there reference to the wives or female relations of these engineers. Her circle of friends and acquaintances and theirs for whatever reason — and it may have been a perceived difference in social class — seem never to have intersected. But Mary Somerville does remark in the manuscript draft of her life that she 'often saw in London' a 'Mrs. Dickson of Felfoot . . . daughter of the Engineer [John] Smeaton who built the Eddistone [sic] Lighthouse'.[49] She and William Somerville met Mrs. Dickson while in the Lake District on their honeymoon in 1812.

Some sentences in her autobiography[50] also give the impression that on this wedding journey she and Dr. Somerville stopped in Birmingham to visit the 'Watt and Boulton manufactory of steam engines at Soho'. She states that 'Mr Boulton showed us everything'. Boulton died in 1809, therefore this visit must have been made not in 1812 after she and Somerville wed but in 1804 when she and Samuel Greig travelled to London following their marriage. It was with Somerville, however, that she was 'escorted by Mr Macintosh of waterproof memory [Charles Macintosh (1766–1843; F.R.S., 1824)]' when they toured 'the manufactories in Glasgow',[51] a visit she also described in her autobiography. The full accounts of machines and of engineering achievements interspersed throughout her memoirs reflect her active interest. Her prediction that man, with all his great engineering capablities, might someday fly even to the moon before the year 1968[52] is amazingly near the mark. Mary Somerville's deep respect for the skilled inventor, the daring engineer, and the sound craftsman and their works is evident in her writings, but men of this sort were not among her close scientific associates.

On the other hand, she was frequently in the company of discoverers and scientific naval men. Her admiration for the Arctic seamen and explorers that sought a northwest passage from the Atlantic to the Pacific was unbounded[53] and coloured by her patriotism and love of the sea. She took great delight in meeting Captain W. E. Parry (1790–1855; F.R.S., 1821)[54] and his assistant, Captain Edward Sabine (1788–1883; F.R.S., 1818)[55] when they returned from their voyage of exploration late in 1820. They became good friends and remained so for the rest of their lives. When Parry set out in 1824 on another search for the northwest passage, Mrs. Somerville made 'a large quantity of marmalade for the voyage' and presented it to him.[56] She prided herself on her skill as a fine cook.[57] Parry gave her mineral specimens and seed of plants brought from Melville Island,[58] and afterward named an Arctic island for her. The ties between the Somervilles and the Sabines also grew stronger as the years passed.[60] Edward Sabine was one of Mary Somerville's chief scientific informants after mid-century.

The Somervilles in this period also became acquainted with Captain John Franklin (1786–1847; F.R.S., 1823)[61] and with his brave second wife Jane (1792–1875). In the next decade they would also see much of the renowned explorer and military man, Captain George Back (1796–1878; F.R.S., 1847).[62] The exploits and discoveries of these men, the firsthand account of their travels and observations, and their writings all whetted Mrs. Somerville's already keen interest in geography.

In this decade the years from 1820 to 1822 were especially happy ones for the Somervilles. They found themselves congenially established in London, pleasantly busy with activities that interested them, and with a host of important friends. Their young daughters were flourishing. Woronzow Greig had done well at Wollaston's old school, Charterhouse, which he had entered in 1820[63] and from which he matriculated to Trinity College, Cambridge in January 1823.[64] James Craig Somerville, after taking a medical degree at Edinburgh in 1820[65] and moving to London the following year,[66] was beginning, after a slow start,[67] to establish the modest private practice that would support him.

Suddenly, three harsh blows struck the family in succession. In June 1823 William Somerville's brother Samuel died in London, where he had sought medical treatment.[68] At about the same time the Somerville's beloved eldest daughter, Margaret, aged ten, became ill and in August she too died. It was a bitter loss to her parents.[69] In the agony of their first grief, they had a further shock: they learned that Samuel Somerville's affairs were in great disarray and that a large sum of money was needed immediately to put them right.[70] They found it, but at great cost to themselves. For a time in the autumn of 1823, the couple seriously considered giving up their house in Hanover Square and moving to Chelsea,[71] but some years seem to have passed before they were finally forced into that unwelcome economy. In 1824 they suffered further financial setbacks from a disastrous Scottish lawsuit and a dishonest Scottish agent.[72]

After the misery and desolation of these months, the solace of foreign travel seemed imperative. In the summer of 1824 the two little Somerville girls were left in the care of the young German governess who had been engaged to teach them that language,[73] just as earlier they had learned French from a French nursery maid. Dr. and Mrs. Somerville, Woronzow Greig, and their old and admired friend, Sir James Mackintosh (1765–1832; F.R.S., 1813), embarked on a tour of the Low Countries and the Rhine. Mary Somerville records that in Brussels she met the astronomer and statistician L. A. J. Quetelet[74] (1796–1874) and at Utrecht the Dutch astronomer Gerard Moll[75] (1785–1838); she continued to be in touch with both these philosophers. In Bonn she renewed her acquaintance with Baron Humboldt.[76]

After their return to London the Somervilles resumed their usual occupations. Mary Somerville, for example, had been for some time in correspondence with Captain Basil Hall (1788–1844; F.R.S., 1816), son of two old friends of the Fairfax family, Sir James and Lady Hall of Edinburgh. Hall had lent journals of his recent voyages to Wollaston and asked him to pass them on to Mrs. Somerville.[77] William Somerville was caught up in the establishment of a new London club, the Athenaeum, designed to attract eminent men of science, literature and the arts as well as men simply friendly to these pursuits. It was exactly to Somerville's taste; for years he busied himself in the admission of candidates.[78]

Their London acquaintance was now very large and they were part of an intellectual society that encompassed distinguished figures in literature, the arts and political life. Mrs. Somerville's accounts in her autobiography[79] of their life in London in these years are not exaggerated, for they are confirmed by the memoirs and letters of others. The couple enjoyed music and the theatre and the excitement of London life but above all they enjoyed the company of their brilliant friends.

3. The First Experimental Paper

The summer of 1825 was exceedingly sunny[80] and Mary Somerville in Chelsea turned to an investigation of a possible relation between sunlight and magnetism, her first systematic piece of experimental work. Simple apparatus was available to her and its use filled some of the long, clear, hot days of the summer. That she should have concerned herself at the time with such a problem is understandable, for studies of magnetic phenomena had recently drawn the attention of several of the philosophers she knew best — Wollaston, Kater, Babbage, Herschel, and Sabine — and of others in the Royal Society.

Oersted's discovery in 1820 of the effect of an electric current on a compass needle gave impetus to a whole series of magnetic investigations. In 1818 not a single paper dealing with magnetism, even magnetism in relation to navigation, appeared in the *Philosophical Transactions*. In 1819 there were two, in 1820, one — but beginning in 1821 the numbers began to rise to a high point of eleven papers in 1826. Mary Somerville, in close touch with burgeoning scientific interests and surrounded by experimentalists, must have felt drawn toward making an investigation of her own.

Philosophers for many years had suspected a connection between light and magnetism but had found it difficult to demonstrate. In 1813 when Sir Humphry Davy was in Rome, he and his young assistant Michael Faraday visited 'the laboratory of Signor [D.P.] Morichini who claimed to be able to magnetize a needle by drawing the violet rays of the solar spectrum along it'.[81] On this occasion Morichini's needle failed to exhibit magnetic properties, and neither Davy nor Faraday were convinced of the validity of the claimed phenomenon. Subsequently two other Continental investigators were unable to repeat Morichini's work successfully. Nature and Mrs. Somerville's circumstances in the summer of 1825 combined to afford her opportunity to make the first study in England of a possible relation between sunlight and magnetism.

She had no information as to how Morichini had carried out his experiments, so had to design her own investigation in its entirety. John Herschel later referred admiringly to '. . . the simple and rational manner in which those experiments were conducted — the absence of needless complication and refinement in their plan, and of unnecessary or costly apparatus in their execution — and perfect freedom from all pretensions or affected embarrassment in their statement'.[82] She made use of materials at hand — a steel sewing needle, a flint glass prism and a lens borrowed from Wollaston, and a variety of common domestic items — paper, coloured glasses and coloured ribbons, wax and the like.

Wollaston appears to have been Mrs. Somerville's mentor in setting up and reporting the experiment. An undated fragment of a letter from him, clearly intended for her, offers advice characteristic of his cautious approach to scientific inquiries, advice that would set her a valuable standard:

. . . and see from the appearance of the transmitted light, that the *red,* the orange & other less refrangible rays are also intercepted. But have you any good reason for thinking that the *chemical* rays are stopped by it?

Your experiments on the effects of coloured lights are so conclusive that I certainly would not couple with them any more doubtful matters or even name Hygrometer Damps or fogs — Whatever conjectures you may have formed on a subject so different, would be far better reserved for your own use in the further prosecution of your inquiries, than even *told* to any one. Doubtful matter dilutes & mars the merits of a good cause. . . .[83]

Not only was the time right for such an investigation, but Mary Somerville's reputation and her connections guaranteed that her study would receive swift and serious consideration. Hers is the first of eleven papers touching on magnetism that appear in the *Philosophical Transactions* in the year 1826 and is the only scientific paper in that journal ever to carry Wiliam Somerville's name in any way. It was he who communicated the work to the Royal Society, where it was read on 2 February 1826.[84]

The work attracted immediate and favourable notice. On 17 February Kater wrote Oersted in some detail of Mrs. Somerville's experiments, commenting that he was 'sure . . . [Oersted would] consider [them] as very interesting' and that he would agree with him 'that such a woman must be considered as an honor to any country'.[85] Kater and others in the Somerville circle had seen a good deal of the Danish philosopher when he visited London in June and July 1823.[86] Herschel also wrote Oersted, declaring that the

. . . only thing very interesting in Science here is the curious series of Experiments made by Mrs. Somerville (I believe you are acquainted with her) on the Magnetism of the violet rays. — According to her, a needle half exposed for a few hours quietly to violet Solar light acquires a permanent magnetism. The exposed end becomes a north pole.[87]

The Somervilles themselves saw to prompt distribution of the paper among Parisian savants. On 14 April Gay-Lussac — who had been in London the previous year — wrote Dr. Somerville that he had received his own copy of Mrs. Somerville's memoir and would dispatch the others to the various recipients the doctor had designated. He added, 'M. de LaPlace m'a chargé particulièrement de faire ses remerciements à Madame Somerville. Je la prie aussi d'agréer les miens avec l'hommage de ma respectueuse admiration'.[88]

Laplace himself wrote Mrs. Somerville on 28 April, expressing his thanks and adding '. . . j'y ai lu avec bien d'interêt, les expériences tres curieuses que vous avez faites sur l'influence magnétique des divers rayons colorés'.[89] The *Bulletin des sciences mathématiques, astronomiques, physiques et chimiques* mentioned her investigation in its April 1826 number.[90] From Rome Domenico Morichini, addressing her as 'My lady Sommerville [*sic*] Chiarissima nelle scienze fisiche', wrote to thank her for having, with her useful corrections and additions, achieved greater and more interesting results than had he.[91]

This notice as well as the high regard given papers in the *Philosophical Transactions* and in the *Bulletin* brought her investigation to the attention of European philosophers at a time when their own interest in the subject was high. In Vienna Professor A. Baumgartner repeated her experiments and obtained similar results which he reported in Arago's *Annales de chimie et de physique*[92] and in the *Zeitschrift für Physik und Mathematik.*[93] Baumgartner's work in turn was reported, with several references to Mrs. Somerville's experiments, in a later number of the *Bulletin.*[94] His researches and hers were accepted as authoritative and valid until Peter Reiss and Ludwig Moser showed three years later that the manifest effect detected was not due to the violet rays.[95]

Many decades later Mrs. Somerville wrote, 'Several unsuccessful attempts had been made in Italy to ascertain whether the most refrangible rays of the solar spectrum possess a magnetic power. I cannot tell why I thought experiments might succeed in England but I made the trial aided by a very delicate test which my kind friend Sir John Herschel made for me'.[96] What Herschel's test was and when it was performed is not specified, but in July 1826, almost half a year after publication of her paper, he and Mary Somerville apparently reconsidered the experiment during a visit the Somervilles made to Slough.[97] After returning to London, Dr. Somerville wrote Herschel asking for a replica for his wife of the 'delicate magnet constructed . . . with a straw' and for the 'Prescription for making the beautiful chrome green' he had shown them.[98] Herschel promptly sent a magnet and some of the chrome, together with the news that he had not succeeded in magnetizing a needle in the 'brilliant sunshine of Friday'.[99]

In the first draft of her autobiography Mrs. Somerville declared:

. . . I imagined I had succeeded or I should not have published my experiment in the Transactions of the Royal Society. I am heartily ashamed of having done so as I fear I may have been mistaken. However I had the presumption to send a copy of it to the Marquis de la Place who good naturedly thanked me for it in a letter he wrote introducing M. Bouvard the astronomer. Since then I have committed all the copies to the flames.[100]

In the printed autobiography the paper is not mentioned. It should not, however, be dismissed. In 1826 the work was considered significant and won favourable notice for Mary Somerville half a decade before her career as a scientific expositor began. At a point when the distinction was an important one, this paper placed her among the scientific practitioners — persons actively engaged in scientific researches — rather than leaving her among the friends and patrons of science. It added immensely to her reputation. At the time no attention was called to the fact that hers was the first paper by a woman — aside from the astronomical observations of Miss Caroline Herschel — to have been presented in her own name to the Royal Society and to have been printed in the *Philosophical Transactions*. It would be two decades before another paper by a woman — and this one also by her[101] — would appear under their banner.

4. Brougham's Commission

In 1824 the French physiologist François Magendie, Laplace's physician, was in London and saw the Somervilles. He carried back to Paris a letter from Mary Somerville to the great philosopher, who sent her a flattering reply and a copy of the fifth edition of his *Système du Monde*. In his letter Laplace observed that he had few readers and judges so enlightened as she. He had just finished the two books of the fifth volume of his *Mécanique céleste*, a task which had forced him to reread

... avec une attention particulière l'ouvrage incomparable des *Principes mathématiques de la philosophie naturelle* de Newton, qui contient le germe de toutes ses recherches. Plus j'ai étudié cet ouvrage plus il m'a paru admirable, en me transportant surtout à l'époque où il a été publié. Mais en même temps que j'ai senti l'élégance de la méthode synthétique suivant laquelle Newton a présenté ses découvertes, j'ai reconnu l'indispensable nécessité de l'analyse pour approfondir les questions bien difficiles qu'il n'a pu qu'effleurer par la synthèse. Je vois avec un grand plaisir vos mathématiciens se livrer maintenant à l'analyse; et je ne doute point qu'en suivant cette méthode avec la sagacité propre à votre nation, ils ne soient conduits à d'importantes découvertes. . . .[102]

Word of this letter and of Laplace's praise of Newton was soon known in the London scientific community. Traces of national feeling still lingered and admiring words by the greatest French philosopher about the greatest English philosopher received general approval and repetition.

Soon after Laplace died in March 1827 Magendie sent Mrs. Somerville a lock of the great mathematician's hair,[103] a sign of high French regard for her talents. By this date she was firmly established in the minds of English and French philosophers as an able student of Laplace's works.

On 27 March 1827 Henry Brougham (1778–1868; F.R.S., 1803) wrote William Somerville a letter which was to change the course of Mary Somerville's life:

... I fear you will think me very daring for the design I have formed against Mrs. Somerville, and still more for making you my advocate with her; through whom I have every hope of prevailing. There will be sent to you a prospectus, rules, and a preliminary treatise of our Society for Diffusing Useful Knowledge, and I assure you I speak without any flattery when I say that of the two subjects which I find it most difficult to see the chance of executing, there is one, which — unless Mrs. Somerville will undertake — none else can, and it must be left undone, though about the most interesting of the whole, I mean an account of the Mécanique Céleste; the other is an account of the Principia, which I have some hopes of at Cambridge. The kind of thing wanted is such a description of that divine work as will both explain to the unlearned the sort of thing it is — the plan, the vast merit, the wonderful truths unfolded or methodized — and the calculus by which all this is accomplished, and will also give a somewhat deeper insight to the uninitiated. Two treatises would do this. No one without trying can conceive how far we may carry ignorant readers into an understanding of the depths of science, and our treatises have about 100 to 800 pages of space each, so that one might give the more popular view, and another the analytical abstracts and illustrations. In England there are not now twenty people who know this great work, except by name; and not a hundred who know it even by name. My firm belief is that Mrs. Somerville could add two cyphers to each of those figures. Will you be my counsel in this suit? Of course our names are concealed, and no one of our council but myself needs to know it. . . .[104]

Mary Somerville had known Brougham and his family in Edinburgh[105] and occasionally the Somervilles and the Broughams saw each other in London.[106] Brougham was a radical leader in parliament, had been one of Queen Caroline's defenders in the notorious 1820 divorce trial, and had a finger in many pies. One of the latest was his Society for the Diffusion of Useful Knowledge,[107] which proposed to bring sound literature and self-improvement within the reach of all by publishing cheap and worthy treatises. Mary Somerville was one of the first persons Brougham approached about preparing a volume. They had carried on mathematical correspondence in the past,[108] and she was well acquainted with the majority of his committee for the Society. The previous year he had launched the new University of London.

In her autobiography Mary Somerville tells how surprised she was at receiving Brougham's request.[109] Her own assessment of her knowledge and acquirements was so modest as to convince her that it would be presumptuous for her to undertake a work of this sort or one on any other subject. Several days after his letter Brougham came himself to Chelsea, where they were living, to urge her to assent, an appeal that had Dr. Somerville's strong support. Together they persuaded her to make the attempt. She stipulated, however, that the undertaking should be kept secret and that, in case of failure, the manuscript should be destroyed.[110] She pointed out to Brougham that a work of this sort, requiring as it did a knowledge of the calculus, could never be popular and that, as 'a preliminary step . . . [she would] have to prove various problems in physical mechanics and astronomy' and include diagrams and figures, which Laplace had omitted. Brougham agreed to all conditions. 'Thus suddenly and unexpectedly', she later wrote, 'the whole character and course of my future life was changed'.[111]

While maintaining her active social and domestic life, Mary Somerville reorganized her daily routine to allow some free time for writing. Her children — their well-being, activities and education — occupied her for a good part of each day, and regular indulgence of her own and Somerville's social proclivities and love of entertainment claimed another large segment of time. By rising early, however, and arranging household matters so as to give herself uninterrupted intervals for writing and especially through acquiring the valuable habit of picking up her pen straightway after any interruption, she was able to make slow progress on the treatise.

Among the visitors to London in these months were the Amici family. William Somerville had been in occasional touch with G. B. Amici since their first meeting in Modena in 1818. Now he had opportunity to introduce the Italian microscopist and his wife to many London friends, including the Katers. The Amicis and their children afterward recollected with warm gratitude the cordial welcome and the many kindnesses they had received from the Somervilles during their English stay.[112] When in the following year the Italian scientist Leopoldo Nobili (1784–1835) planned a brief visit to London, Amici wrote Somerville asking him to introduce Nobili to English scientists, including Mrs. Somerville, whom he declared known already to the electrician by reputation and admired by him for her many 'enviable virtues'.[113]

In the summer of 1827 she remained at home while Dr. Somerville went abroad for a short restorative holiday[114] after a bout of ague. Woronzow Greig was also in Europe. He had taken his degree at Cambridge earlier in the year[116] and was now abroad on a grand tour with his close friend and classmate, William King, later Lord Lovelace. On his travels Greig collected mineral specimens and information for his mother, and in Rome was much in the company of Charles Babbage, abroad for a year after a series of family bereavements.[117] While not himself a scientist, the young man found much in common with his mother's scientific friends and was warmly regarded by them.

William Somerville was back in England in time to take his wife to Beechwood for a house party with Wollaston and other guests in September 1827.[118] The financial gloom that had hung over the Somerville family since mid-1823 and that had forced their removal to the isolation of Chelsea lightened enough for them to contemplate a season in the metropolis. Since they had given up the Hanover Square establishment, they rented a small house at No. 6 Curzon Street,[119] next door to their old friends the Misses Berry. Mary Berry (1763–1852), the well-known authoress, and her sister Agnes lived at No. 8 Curzon Street, a house owned by their cousin, Robert Ferguson of Raith. There, for more than half a century, the Misses Berry's salon was 'the resort of the best company in London'.[120] Once again the Somervilles were at the center of London scientific, literary and political life.

As it chanced, this was the last London season in which Mary Somerville would be associated with the giants of her early London scientific circle. In December of the following year, her chief mentor and dear friend, William Hyde Wollaston, died after some months of illness. His mind remained clear to the end, and a fortnight before his death he dictated the following message to her:

Dr. Wollaston wishes much, that under his present sufferings he could express to Mrs. Somerville in adequate terms, how much & how frequently he has enjoyed her society & conversation, and how ardently he wishes that she would retain as a token of friendship the cabinet of models of crystals, made by Larkins, which he formerly lent her; he hopes Mrs. Somerville will keep it for the purpose of instructing her children in the useful Science of Crystallography.

Dorset Street
the 7th. Dec[r.] 1828[121]

Twenty years later, her own children grown, she presented the cabinet to the John Herschels for the use of their progeny.[122]

When, at Wollaston's direction, a post-mortem was performed, young James Craig Somerville assisted the two eminent surgeons, William Babington

(1756–1833; F.R.S., 1805) and Benjamin C. Brodie (1783–1862; F.R.S., 1810), who carried it out.[123] The Somervilles, together with Henry Warburton, had become Wollaston's most intimate friends, and it was to them that other friends frequently addressed letters of inquiry about his condition during his last illness.[124, 125] Surviving among the Somerville papers is a letter from the author William Sotheby (1757–1833; F.R.S., 1794), long associated with Wollaston in the Royal Society, expressing concern for Mary Somerville after 'her bitter solicitudes' of the past weeks and deploring the loss that Wollaston's passing inflicted on 'the intimate circle of Friendship . . . [and] the world of Science'.[126] Contrary to an often repeated story, Wollaston's papers were never placed in Mrs. Somerville's charge.[127]

A few months later, on 10 May 1829, Thomas Young died. The last letter from him in the Somerville papers is a lighthearted apology to Mrs. Somerville for 'wrongfully' suspecting that she had one of his books when in fact it was on his own shelf.[128] After Young's death, only the ailing Kater and the amateurs of science — Blake, Warburton, Chantrey and Sebright — remained of the Somervilles' first London scientific circle.

The last winter in which it was intact — that of 1827–28 — had, however, its share of excitement for the group. Several of them were closely involved in the 30 November election of a new President of the Royal Society. Sir Humphry Davy, forced by ill health to resign the Presidency on 7 November, wished to see installed in his place Robert Peel, a Fellow since 1822 and a gifted mathematician[129] as well as a skilled and farsighted Tory politician. A reform group in the Council, led by Wollaston, protested. Their objection was not to Peel personally, although many of them disagreed with him politically, but rather to the nomination of his successor by a departing P.R.S. Peel, returning to government shortly before the election day, withdrew his name and the interim President, Davies Gilbert (1767–1839; F.R.S., 1816), became the uneasy choice of the Society. From this time forward, however the reform group grew more and more insistent on change.

Its leadership, after Wollaston's death, passed to some of the younger men of his circle, principally to Herschel and Babbage, who were already closely associated with the Somervilles. By 1826, for example, John Herschel and they were exchanging frequent letters and visits.[130] Charles Babbage, after his return to England in 1828, was also often in their company.[131] In the coming decades these friendships, already strong, would deepen.

Mary Somerville's horizons too were widening and changing. In the drafts of her autobiography she observes:

Though still occasionally occupied with the structure of the mineral productions of the earth, I became more interested in the formation of the earth itself. Geologists had excited public attention and shocked the clergy and the saints by proving beyond a doubt, that the seven days of creation is an eastern myth of enormous geological periods. nevertheless the contest was even more keen than at the present time [1869] with the pre-Adamites. It lasted long too. Our friend Dr. Buckland comitted [sic] himself by taking the clerical view in his Bridgewater Treatise [1836] but facts are such stubborn things that he was obliged to join the geologists at last.[132]

Captain R. I. Murchison, whom she had met in Rome in 1817 was embarking on the same sort of study in London; young Charles Lyell had settled in the capital and was devoting his time to geology. The Somervilles saw more and more of their geological friends. In February 1829, they went with the Murchisons to Oxford to spend a glorious week as the guests of William and Mary Buckland in Christ Church.[133] Mary Somerville had much in common with these two scientific wives, Charlotte Murchison and Mary Buckland. All became lifelong friends. Because of her continuing support of the new geology, Mrs. Somerville was denounced in the next decade from the pulpit of York Minster by the polemical Reverend William Cockburn (1773–1858), who mounted a long-running attack against the 'new geologists'.[134, 135, 136]

By 1829 word of her Laplace project was beginning to reach the scientific community. The Somervilles were in Scotland in the summer of 1829 and saw David Brewster. He described Mrs. Somerville to his young pupil James David Forbes as

. . . certainly the most extraordinary woman in Europe — a Mathematician of the very first rank, with the gentleness of a woman, and all the simplicity of a child. She has nearly finished a work in 3 vols — An illustration of the Mécanique Céleste of Laplace, She shewed me some parts of ye Mss and I have no doubt it will gain her the highest reputation. She is also a great Natural Philosopher and a great Mineralogist.[137]

By the end of the decade the training which Wollaston, Young, Kater and others had given Mary Somerville had produced another British philosopher, 'the most extraordinary woman in Europe'.

CHAPTER 5

THE MECHANISM OF THE HEAVENS

1. The Atmosphere of 1830

The year 1830 is significant in the cultivation of science in Great Britain not because of any great scientific discoveries or innovations but because of human factors. It was a year of intense self-examination in the English Scientific Establishment and of the bitterest confrontation yet in the Royal Society between those insistent on reform and those determined to maintain traditional ways. It brought to the fore two new figures — King William IV and his younger brother, the Duke of Sussex — who would have roles in the functioning of British science in the coming decade. It marked a point in the history of the Royal Society when, as reformers were defeated, the reforms they advocated slowly began to be instituted. All these events took place against a background of change and alarm in the civil sphere that had a profound influence on activities in the scientific sphere.

The decision to publish Mary Somerville's first book and the reception subsequently accorded it were closely tied to these matters, for they created an atmosphere favourable to the work and to its publication. The two most important occurrences in this regard were the appearance of Charles Babbage's *Reflections on the Decline of Science in England and on Some of its Causes*[1] in May 1830 and the defeat of John Herschel for the presidency of the Royal Society in the following November. Both these happenings had their roots in the recent past and both involved a number of persons important in the Somervilles' scientific set. In neither of them was Mary Somerville an active participant; in the second of the two her husband played a minor role. Yet the feelings and misgivings these events engendered contributed substantially to the success of her first book.

Charles Babbage's attachment to science was unquestioned, and his views as to how English science might best be advanced well formulated by 1829. Ever since his days at Cambridge he had urged change: adoption of the new French analysis, innovative use of scientific theory and findings (as in his calculating engine), increased state support and recognition of the scientific enterprise, reforms in the functions and management of scientific bodies, and, above all, professionalization of science. In some quarters, he succeeded in gaining at least limited support, in others he met with indifference.

Particularly galling to Babbage and many other scientific fellows of the Royal Society was the steady increase in the number of non-scientific fellows entering that body. Indiscriminate admission had by 1827 swelled the fellowship to almost 700, the vast majority of whom took no part whatsoever in the scientific business of the Society. Early in that year the Council, on the urging of James South (1785–1867; F.R.S., 1821), appointed a committee to consider how best to limit membership in the Society and to make suggestions conducive to its welfare.[2] Seven of the eight members of the committee — Wollaston, Kater, Herschel, Young, Babbage, South and Francis Beaufort, all of whom were part of the Somerville circle — were reformers and the eighth, Davies Gilbert, a traditionalist. The committee report, which Gilbert signed with unvoiced reservations, recommended that membership should be fixed at 400, that the names of candidates be circulated in advance to all fellows, that presidential power of nomination be curbed, and that a standing committee on Society finances be established. Since the report came late in the spring, its consideration by the Council was postponed until autumn. When meetings resumed, however, the Council and the Society were caught up in choosing a successor to Sir Humphry Davy. Chiefly at issue was the prerogative of a retiring president to name his successor, but there were also reminders of the 1820 debate over the selection of a practising scientist or an influential friend of science as head. In the end Davies Gilbert was a choice of convenience, for Davy's candidate, Robert Peel (1788–1850; F.R.S., 1822),[21] withdrew and the reformers failed to find a scientific nominee. In the transition, action on the 1827 reform report was postponed.

Babbage in his stay abroad from late 1827 through 1828 observed foreign science and technology at first hand and met many of the leading scientists in Europe. At Berlin he attended the annual gathering (since 1822) of German philosophers — the *Deutscher Naturforscher Versammlung* — and was impressed by the lavish reception the Prussian government gave the group and the professionalism the organization displayed. He returned to England convinced that the superiority Britain had enjoyed in science since the days of Newton was seriously threatened. He returned also to the vexing problem of gaining more government support for construction of his calculating engine. Late in the spring of 1829, the Prime Minister, the Duke of Wellington, recommended that an additional £3000 go to the project, but, to Babbage's despair, tedious discussions and negotiations with unsympathetic officials delayed payment until February 1830.

Babbage returned to find also that the Royal Society had not acted on the 1827 committee report. Gilbert, re-elected in 1828 in the absence of any other acceptable candidate, had always privately viewed the measures proposed as dangerously democratic. An experienced politician well aware of the occasional usefulness of doing nothing, he — as head of the Society — avoided an end he found abhorrent simply by letting the report lie. His failure to act on these proposed reforms became a festering sore in the Council and in the fellowship.

Another source of dissatisfaction was the *Nautical Almanac* and the operation of the Royal Observatory at Greenwich, both matters of primary concern to a seafaring nation and into which the Royal Society had been drawn through its advisory capacity. Thomas Young, whose appointment to the Board of Longitude Babbage protested in 1818 on the grounds that Young had no practical knowledge of astronomy, was in charge of the *Almanac* and even before his death in May 1829 was under fire for its inadequacies. Francis

Baily's attack on Young's superintendence[3] had quickly been followed by two pamphlets on the subject by James South.[4, 5] Young's death quietened agitation for a time, but Baily, South and Babbage came together in their determination to reform the *Nautical Almanac* and to be watchful in all matters relating to it and to the Royal Observatory.[6]

The period from December 1828 to June 1829 was a devastating one for the British Scientific Establishment. Within six months of each other, three giants of British science — Wollaston, Davy, and Young — died. David Brewster, in a comment to Oersted, reflected not only the feeling of scientists but of Britain: 'What a loss we have sustained . . . it has no parallel in the History of Science'.[7] Change was being thrust upon British science and Babbage and the reformers were determined that it should be along paths they had long advocated.

In the autumn of 1829 Babbage began a private investigation of the management and functions of the Royal Society over the last decade. His findings certained him in his conviction that science in Britain was neglected and declining. In May 1830 he published a polemical work, *Reflections on the Decline of Science in England and on Some of Its Causes,*[8] which not only described the shortcomings he detected, particularly in the Royal Society, but offered remedies for them. The author quoted extracts from Thomas Thomson's piece on chemistry in the *Encyclopaedia Metropolitana,* from Davy's *Consolations in Travel,* and from John Herschel's recent piece on 'Sound', also in the *Metropolitana,*[9] to show that he was not alone in his gloomy assessment. He denied any 'hostility to the Royal Society' but did condemn 'mismanagement' in it,[10] making his case through extracts from minutes of its proceedings that he had obtained without Council permission. The book attracted wide attention in the scientific community and caused a furore in the Council and in the Society that went on for the next four months. There were demands that Babbage be expelled and there were stout defences of his actions. Disputes in Council were generally kept private, but this one spilled over into the public press. A number of letters and comments about the decline of science and the Royal Society appeared in the *Times* in May, June and July 1830.[11]

That newspaper, sympathetic to Babbage and the reformers, had taken early notice of his book, calling it 'a work which deserves and must infallibly attract, much attention'.[12] Another important organ, the *Athenaeum,* a weekly with a strong liberal bent and ties to Cambridge, devoted its lead article of 22 May to the book[13] and continued to take notice of the ensuing controversy. Its rival, the *Literary Gazette,* in what must be considered a deliberate act, mentioned it only in a list of new books.[14] Neither the *Edinburgh Review* nor the *Foreign Quarterly Review* — both given to lengthy notice of scientific works — nor the Benthamite *Westminster Review* took any heed of Babbage's bombshell, but in Edinburgh David Brewster, a man also disenchanted with the Scientific Establishment, wrote a long and favourable review. It appeared anonymously in the October issue of the *Quarterly Review.*[15]

The scientific world and a good part of the seriously literate public, therefore, were aware of a possible decline in British science almost a year before Mary Somerville's work appeared. Among a considerable number there was concern and dismay. They looked for evidence to counter the charge, and Mrs. Somerville's book came to hand as an easy refutation and reassurance. Babbage's widely publicized polemic influenced in this way the initial response to her work.

Another unexpected factor in its reception was the dejection felt by the reform element in British science after the Royal Society election of 1830. In John Herschel the reformers had their strongest candidate; they had worked hard for his election and been confident of victory; in the end the forces of tradition had bested them by a narrow margin. The shock of this defeat set their leaders on a determined search for means other than the Royal Society through which to bring about the changes they viewed as pressingly imperative. Mary Somerville's book would have been received kindly by these leaders under any circumstances — for they were gentlemen — but their failure to assume direction of the Royal Society gave an added impetus to her work. It was an admirable example of what they had been advocating, to be praised not only for its own merit but for the talent and encouragement that it manifested.

In the election of 1830 the issue once more was whether the Society should be headed by a working scientist or by — in this case a very recent — friend of science. Once again, as in 1820, the election took place against a background of change and disturbance in the civil sphere. On the night of 25–26 June King George IV died at Windsor and his brother, the Duke of Clarence, succeeded as William IV. The new king in the seven years of his reign would find himself on occasion part of the changing pattern of relations between science and the state. A goodhearted, informal man, economical and unabashedly patriotic, 'extraordinary only in his very ordinariness',[16] the new monarch was more adaptable to change than his predecessors. He entered on his reign with zest. Acting on the advice of Sir Robert Peel, he promptly displayed a friendly disposition toward British learning and science. Within two months of coming to the throne, he graciously placed himself at the head of the newly-fashioned Geographical Society (which thus joined the Royal Society and the Royal Society of Literature as learned bodies under the King's sponsorship). He also declared interest in King's College under construction for the new University of London and funded for the first time the Royal Medals established in 1826 by his late brother. Early in July, he carried out his brother's intention to bestow a baronetcy and pension on James South.

South, a dangerously intemperate man whose large fortune, acquired by marriage, permitted him to maintain private observatories in London and in Paris, had become so irked by the course of astronomical reform in England that he began toying with the idea of removing permanently to Paris. According to a story which appeared in the London *Times* of 5 August 1830, he had actually written to the French government and received their cordial assent to his proposition. In early July a title and £300 annually 'to be applied by him to the promotion of astronomy'[17] persuaded him to remain in England.

Even without these inducements, South might well have altered his plans, for on 26 July revolution broke out in Paris. The despotic Charles X was forced to flee, and on 9 August Louis-Phillipe was installed as king by victorious republican forces. Before the end of August a successful uprising took place in Brussels. Before the end of the year the Poles were in revolt against the Russians; liberal factions in Spain and Portugal rose against their tyrannical governments. Reports of the Paris insurrection did not reach English newspapers until 3 August, by which time polling in the general election following the change in monarchy was well underway. Although too late to have much influence on the outcome, the news did arouse great popular excitement and moved the Whigs — in the majority for the first time in decades— to public announcement of their support of parliamentary reform. To many, especially among the Tories, the dangers apprehended five decades earlier with the first French revolution seemed to threaten again. Reform was in the air and to the fearful it loomed as the entering wedge of disaster. Their alarms would intensify in the autumn.

Another member of the Royal family, one of the King's younger brothers, Augustus Frederick, Duke of Sussex (1773–1843; F.R.S., 1828) would be a central character in the drama at the Royal Society in October and November 1830. A large, tall, cheerful man plagued by recurring ill health but ever on the lookout for useful employment, Sussex was regarded by many as a bit of a buffoon. He had long occupied himself with charities and with assembling a large and varied library and was already President of the Society of Arts and Grand Master of the Freemasons. He enjoyed a reputation for kindness, generosity and independence and was a 'fluent and sound . . . [if not] brilliant speaker'.[18] In him Davies Gilbert saw his ideal P.R.S.: a gentleman of great position, influence and fortune who would preside with dignity over the venerable body, enhance through his own standing the reputation of the Society, and advance the cause of science through his own station in life and his whole-hearted devotion to the national good. The fact that Sussex was not a natural philosopher and had shown little interest in the subject was of no consequence.

When Gilbert accepted the presidency of the Society in 1827, he did so with the intention of stepping aside in a year for the Duke,[19] whom he and Davy had approached informally on the subject in 1826. At that time Sussex gave the offer no serious consideration, since election would inconvenience him by requiring his presence in London every St. Andrew's Day.[20] After his physician, librarian and friend, Dr. T. J. Pettigrew (1791–1865; F.R.S., 1827), became a fellow in February 1827, the Duke began to regard the suggestion — repeated privately and explicitly to him in October 1827[21] —more favourably. In May 1828 Gilbert in an almost surreptitious fashion had Sussex elected a Fellow and hoped, even though H.R.H. did not meet the statutory requirement of being on the Council, to make him president in November. The Duke was now willing, but a strong segment of the Council and fellowship —some disturbed by the irregularity and others unhappy also at the prospect of a non-scientist — 'objected on the grounds that the Society would have lost some of its independence by being tied to the Royal Family'.[22] Sussex, recognizing the internal tensions that prompted this reaction, withdrew temporarily.

Gilbert, re-elected in 1829, did not abandon his intention of relinquishing the office to the Duke nor waver in his conviction that a president had the power to choose his successor. After the publication of Babbage's book and during the furore which followed it, he maintained in public a dignified silence while joining in private in the open and often blunt discussion at the Council. Attendance at the weekly sessions of the Royal Society in May and June was exceptionally full in anticipation by the Fellows of outright confrontation between the P.R.S. and his chief critic Babbage. The two men displaying remarkably good sense and restraint, met the situation with a combination of conciliatory words and judicious occasional absence that preserved a facade of civility.

Gilbert, although often irresolute and hasty and much given to vagueness, was not despotic, vindicative or distrustful. By early July, however, he began to realize the depth of dissatisfaction over his performance as P.R.S. and the extent of antipathy to his traditionalist view of the Royal Society. He resolved to go but was determined, as an act of service to the Society, to see Sussex installed in his place. His ultimate success came only after three months of bitter manoeuvering.

In mid-August watchful reformers found a premature hint of Gilbert's plan in a piece[23] in the *Literary Gazette* by its editor, William Jerdan (1783–1865), known to be sympathetic to Gilbert and to the Duke. But all London was on holiday and no concerted response was made. Even Gilbert himself did nothing until 10 September when he wrote Dr. Pettigrew from Cornwall that, while still holding in the abstract to the view of the 1828 Council about the necessity of preserving the Society's independence, he was now, because of the direction of current events and opinions, 'practically most desirous of seeing the Society under the protection of one nearly connected with the throne'.[24] Before the end of the month the Duke had accepted Gilbert's offer of the presidency,[25] and Gilbert and Pettigrew began a correspondence regarding nominations for other officers and the Council.[26] Not until 1 October did Gilbert inform the members of the Council, by letter, of his intention to step aside for the Duke.[27]

Almost simultaneously the press began to take an interest in the matter and over the next two months, as tension and excitement mounted, the Royal Society was much in the public eye. Regular reports appeared in the newspapers, along with frequent unsigned 'letters to the editor'. The weeklies devoted more space than usual to doings at Burlington House, and by late November the *Literary Gazette* used the term 'Row-all Society'.[28] Two inflamatory pamphlets by James South[29, 30] attacked Gilbert and the traditionalists, while a third — by A. B. Granville (1783–1872; F.R.S., 1817)[31] — supported the Duke of Sussex. In private, furious discussion, exchange of letters and informal and formal meetings multiplied. The lines were quickly drawn: on one side the reformers, determined to put an end to presidential nomination of officers, to maintain the Society's independence, and to install a working scientist as its head. On the other, the traditionalists joined to preserve the old patterns, to resist the encroachment of democracy in Society affairs, and to place in the chair for the first time a prince of the blood.

Their numbers were swelled by events in the outside world. Already shaken by revolution abroad in the summer of 1830, Britons found themselves in the autumn in the midst of uprisings at home. A third calamitous harvest in as many years led to sporadic and unorganized violence in some of the southern agricultural districts and to strikes in turbulent northern towns suffering from rising food prices, shortages, and declining employment. From late October until early December the *Times* and other newspapers carried long and lurid accounts of disorders and incendiarism, reports of unverified rumours, details of apprehension and savage punishment of rebellious farm workers and strikers. Gentlemen and tradesmen in the provincial towns and outlying districts organized themselves into patrols ready to put down any hint of violence. Many of the Fellows of the Royal Society were large landowners, often magistrates in the restive counties; a few actually had their own hayricks burnt. Scientific circles were conversant not only with the daily rumours and frightening reports in the papers but heard at first hand of mobs, arson, violence and the ensuing repression and punishment. By the middle of November troops were called out, riots were expected everywhere, funds fell, and the Duke of Wellington's life was reported in danger. To many the old ways and the traditional approaches suddenly took on added attraction. In the end, this civil unrest would stir many fellows who generally did not bother to vote, to do so — and for the Duke of Sussex.

The Somervilles' sympathies since their first close association with London science in 1820 had been with its reform element. Since many of their intimates were on the Council and active in the strong press for reform, the two were well aware of events, stands and attitudes in the Royal Society. They knew that their friend John Herschel had, since his resignation as secretary in 1827, taken little part in Society affairs and had avoided factional disputes. Confident that reform must ultimately come and hopeful that it could be accomplished without damage to the fabric of the Royal Society, he busied himself with his scientific work, his writing, service on scientific committees, and, since his marriage in 1829, his family. Although he fully understood his friend Babbage's frustrations and agreed with him about the need for reform, Herschel deplored the caustic tone of the *Reflections*[32] and remained aloof from the controversies of the spring and summer. When in late June Gilbert made overtures to him through the offer of a commission to write the Bridgewater treatise on astronomy, he curtly refused in a letter hinting at some of Babbage's charges about the lack of support of worthy professional scientists. He suggested that the deserving William Ritchie (1790–1837; F.R.S., 1828) be chosen in his place, as the £1000 'windfall'[33] accompanying the commission would be of greater assistance to that struggling engineer (later professor at the University of London, 1832–1837) than to him.

To the reformers, nevertheless, Herschel appeared indisputably the best choice for P.R.S. Recognized as the outstanding man of British science, fully committed to the views held by the reformers, conversant with every aspect of the Royal Society, personable and honourable, well fitted to the office in terms of character, accomplishments and heritage, Herschel at age 38 would bring vigour, intelligence and skill to the presidency. Not until after mid-October, however, were his friends able to persuade him that he must intervene. A piece in the *Literary Gazette* so aroused his anger and disgust that he at first considered resigning from the Royal Society. The journal reported that differences in the Council were in the process of being adjusted, that all obstacles to Sussex's unanimous election seemed to be vanishing, and that several designated reforms would be adopted.[34] The exchange implied — implementation of reforms long adopted by the Council and institution of a dubious new organizational structure in return for election of the Duke — and its mode of transmission — in a public journal — infuriated the reformers and brought Herschel actively into the fray. With him came William Somerville, to play his own modest part.

Before the meeting of the Council on 4 November — the first since 10 July — the reform forces had rallied. Thirty-two Fellows — none of them on the Council and most of them members also of the Astronomical and Geological Societies — from a sense of 'duty' signed a paper requesting the President and the Council to communicate 'to the Society at large' the nature and particulars of any correspondence that may have taken place' on the matter of changing Society officers.[35] For weeks there had been talk of a 'Pettigrew Correspondence' and a 'Pettigrew cabal', of letters privately exchanged that made promises and conditions involving the Society and its officers.[36] William Somerville, who generally avoided controversy, was the last Fellow to sign the requisition. The names of the Somervilles' particular friends Herschel, Baily, Fitton, Broderip, Beaufort, Brodie, Horner, Murchison, Whewell and Chantrey were already on it.[37]

The 'Requisitionists' expected 'either the publication of the entire correspondence . . . or the calling of a special meeting of the Society, before which it might be read',[38] but the Council, 'unacquainted with the particulars of the negotiation . . . referred to' called upon subscribers to the requisition to meet with the President and Council, and form with them 'a committee to consider the proper measures to be taken . . . .'.[39] The meeting was fixed for 11 November.

It took place at Burlington House in the afternoon before the regular Council meeting. Gilbert, after making an initial statement, tactfully withdrew and Henry Warburton was elected to preside. Two of the 'Requisitionists' — H. T. Colebrook (1765–1837; F.R.S., 1816) and W. H. Fitton (1780–1861; F.R.S., 1815) — introduced a resolution, unanimously adopted, directing 'that the Officers and Council be selected from among such members of the Society as are by their acquaintance with the conditions and interests of science best qualified to discharge such offices',[40] i.e., practising scientists. The Council was to — and subsequently did — devise a list of such Fellows and circulate it among the membership for their vote on 30 November. In the discussions there was no dissent from the opinion that 'a Prince of the Royal Family would not . . . be an eligible President of the Society'.[41] Warburton undertook to make these views known to H.R.H.; it was widely expected that the Duke would withdraw from consideration.

To Gilbert, who had long equated reform with revolution, these actions were not only a shattering personal blow but spelled, through democratization, a total change in the character of the Society that would 'reduce it to a level with the minor societies' [42] such as the Geological. To add to his distress, the Tory government — entrenched for decades in Westminster — toppled five days later in the winds of reform. On 17 November Lord Grey became Prime Minister, the Whigs took over the offices of government, and their promised parliamentary assault was underway. In the face of these menacing pressures, Gilbert moved to salvage as much of his establishment as he could and to guide the Society with dignity toward the Anniversary Meeting a fortnight away.

At the next Council meeting — on 18 November — forty names were chosen by ballot to be submitted to the membership for selection of the new Council. The names of Gilbert and Babbage were among them, along with those of Herschel and fourteen other Requisitionists; those of Sussex and Sir James South — whose polemics had alienated many — were missing. Gilbert did succeed on one point: recision of that part of the earlier resolution referring to 'Officers of the Society', which in effect made the candidature of the Duke possible even in the absence of his name from the list. Gilbert himself had great fear that H.R.H., who to this point had stood firm, might, in the face of all this unpleasantness, withdraw.

Press notice of happenings at the Royal Society throughout October and November had been full and as St. Andrew's Day approached it was intensified. On 23 November the *Times* printed a denial — attributed to the *Athenaeum* — of any intention on the Duke's part to retire and declared that up to the moment he was the only candidate.[43] The reformers were now fully on notice — a week before the election — that there would be a battle.

They still had no candidate. The names of Warburton and South were occasionally heard[44] but Herschel was the man they wanted. His friends pressed him hard; Murchison even put together a formal request from a group of scientific fellows, mainly from Cambridge.[45] At the last moment a reluctant Herschel agreed. If those who shared his conviction that the Society should be headed by a practising scientist, that its presidency was not to be passed like a bauble or sceptre from one amateur to another, that its procedures and management must be reformed to enable the body to serve science and the nation nobly, and that its role was greater than mere provision of entertainment for dilettantes — if fellows of this mind wished him to be their standard-bearer — then he must, he felt, at this late hour accede. He did so sadly, writing Babbage on 26 November[46] and again on the 27th,[47] that he did not want the office, had no panacea for the ills of the Society, and, if elected, would serve only one year.

Herschel himself did no more than agree to stand. Having done so, he made no declaration of policy, issued no appeals on behalf of reform or himself. His perception of the problems of the Royal Society and of science and any plans he might have to remedy them were never publicly enunciated. All public statements came from others — Babbage and South most loudly

— and without his authorization. His support came primarily from the reform element in the Society but there were other ties also that brought him votes: school and university loyalties, shared interest in astronomy, mathematics and physical science, fellowship in specialist societies, opposition to royalty in the chair, and often the simple conviction that the head of science should be a man of science.

The Society met on 25 November for the last time before the election. The *Times* that morning carried an unprecedented advertisement:

> The undersigned Fellows of the Royal Society being of the opinion that Mr. Herschel, by his varied and profound knowledge and high personal character, is eminently qualified to fill the office of President, and that his appointment to the chair of the Society would be peculiarly acceptable to men of science in this and foreign countries, intend to put him in nomination on the ensuing election.[48]

Sixty-three names in alphabetical order were affixed. One of them was that of William Somerville.

Not only the method but the action itself was extraordinary. A paid advertisement in the public press on what had for almost two centuries been a private matter among gentlemen! Opposition to one of the Royal blood over an office that the majority considered merely honorific! The reformers hoped, through this unusual public notice, to inform the Duke 'correctly of the previous circumstances, and of what was in progress'[49] and believed that he would, under these circumstances, withdraw from a contest scheduled to take place in five days. He did not. On 29 November[50] the advertisement appeared again, this time with seventy-seven signatures; Charles Babbage was among those who had added his name. And on the morning of the election, 30 November,[51] it was inserted for a third time, now with eighty names affixed.[52] A swell of Herschel support was becoming manifest. The Duke, it was reliably reported, took the advertisement 'as a personal insult' and declared that he would neither withdraw nor meet 'in Society or . . . [sit] down at table, with the individuals who had signed it'.[53]

Up to the morning of the election Babbage and his faction hoped the Duke would withdraw.[54] They were well aware that Pettigrew and other Sussex partisans had been canvassing actively and that the Duke enjoyed an immense advantage because of his birth and position. The very state of the nation, with farm-hands in rebellion and factory hands on strike, propelled the timorous to his support. There was talk of 'bullying letters . . . [and] unfair influence' being exerted to restrain Herschel support, hints of ducal retribution.[55] The reformers never wavered; their faith lay in the overwhelming logic and persuasiveness of their case and in the growing support they saw for Herschel. While scorning the employment of 'indirect expedient[s]',[56] they neglected no direct appeal. Personal letters went from Herschel supporters in London to those outside the metropolis urging attendance at the 30 November meeting and a vote for the science candidate.[57] For some, however, the press of outside events loomed larger than the necessity of coming to London. Buckland wrote from Oxford that he would do so

> . . . if it be a very hard-run thing . . . but I shall be very sorry to leave home . . . for my wife's father and mother, six miles from here, are in hourly expectation of a mob from Abingdon to set fire to their premises, and there are threats of a mob coming into Oxford from the neighbourhood of Benson, and our streets, every night, are on the point of a row between town and gown. . . .[58]

In the end so bright did the prospects of the scientific party appear and so confident did Babbage, Fitton and Beaufort — the three managers of Herschel's campaign — become that they 'at the last moment relieved a number of his supporters from coming to London to cast their votes'.[59]

Their optimism was ill-founded. An 'unprecedently numerous'[60] attendance marked the Anniversary Meeting, the 'rooms were excessively crowded',[61] more so than at any time in the past three decades.[62] The *Times* estimated 'not less than 500 . . . of the 720 fellows resident in England' were there.[63] Each faction had its slate of Council candidates and balloting took place 'with unparalleled ardour on both sides'.[64] In a penetrating account of the meeting, the *New Monthly Magazine* referred to three parties as being involved:

> . . . the Reformers . . . the Ultras . . . [and the] neuters . . . [the first consisting] of those members who are more especially attached to the cultivation of mathematical science and its auxiliary branches, and who, with great propriety, had nominated Mr. Herschell, a gentleman eminently qualified, from his various attainments, to do honour to the chair . . . Another party . . . the Ultras, or old government party, nominated a member of the blood royal for candidate. While no inconsiderable number of the Fellows preferred perhaps the wiser or at least the safer course, of looking on, and remaining neuter at the proceedings of the contending parties.[65]

The Somervilles in this instance did not take the safer course but publicly and from the first supported Herschel. William Somerville's action in being one of the Requisitionists and in signing all three *Times* advertisements was in some respects unusual for a man who throughout his life sought the favour of the powerful and well placed. On this occasion, friendship, the gratitude that he and his wife owed Herschel, their long association with the reform group, and the strong possibility of its being soon the faction in power led him to act decisively.

The Council election was crucial, for the statutes required that the P.R.S. be chosen from among the Councillors. Sussex headed the slate put forward by the traditionalists, Herschel that of the reformists. Of the 33 nominees, nine had the endorsement of both parties and all nine were elected. Among them was young J. W. Lubbock, who had been at Trinity College with Woronzow Greig and had become a friend of the Somervilles; he succeeded Kater as Treasurer.

Whether out of wariness or out of embarrassment or disgust at the prolonged public controversy, only a third of the total fellowship voted, 230 out of 687. At 3 p.m. one of the scrutators, Thomas Amyot (1775–1850; F.R.S., 1824), declared the Duke of Sussex nominated by a vote of 119 to 111.[66] The reformers had expected to win; they 'knew the strength of the Royalist party [but] they did not calculate the number of neutral members'.[67] Herschel's vote was little more than half the total of scientific fellows in the Society, Sussex's less than a quarter of the non-scientific, yet the narrow difference of eight was in favour of H.R.H.

Gilbert on the evening of the election gave forty or fifty of the fellows dinner at the Crown and Anchor Tavern. His sister, Mrs. John Guillemard, told Maria Edgeworth shortly afterward that '. . . her brother was delighted to have done with the Presidency of the Royal Society . . . and pleased all . . . by his way of leave taking and his wish to leave all scientific and other dissensions where they should be left — in oblivion'.[68] Once he had left the office, there was no ill feeling against Gilbert. He and the Somervilles remained on the best of terms until his death in 1839.

Among the Herschel supporters there was profound initial shock and dismay. Gideon Mantell voiced their first reaction and their first hope in his journal entry for 1 December 1830: 'The Duke of Sussex elected Pres. R.S. but only by a majority of seven! — Surely his R.H. will not accept it!'[69] On the same morning the *Times* reported, without authority, hints 'that the Duke of Sussex would probably resign the invidious honour acquired under such singular circumstances'.[70]

It soon became evident, however, that the Duke had no intention of resigning, and the majority of Fellows were weary of the fray. Maria Edgeworth expressed the view of most of the Duke's supporters and opponents when she wrote a friend on 8 December:

Next to the change of ministry and politics the next subjects of the day have been the election of the President of the Royal Society and the expected trial of Mr. [St. John] Long [for medical malpractice]. Mr. Herschel had I believe all the wishes in his favor of every man of science in the Royal Society even of those who from circumstances were forced to vote in favor of the Kings brother. Herschel has been happy in having all the honor and escaping all the trouble. It would not have been a situation suited either to his fortune, taste or manner . . . It is my secret opinion that it is best suited as it is. Tho the Duke of Sussex is not a man of science he may do the honors to men of science and fill the Presidents place more agreeably than a rival or a judge. How he is to get through his speeches or discourses to the Society annually is another affair but bought science I suppose may be as good as bought wit and if he reads well or can get it by rote — alls well . . . Poor Davies Gilbert to whom the place was in every way unsuited is well out of it. I hope he thinks so.[71]

Press comment, which had become increasingly vociferous as the battle intensified, showed restrained relief at the outcome. The *Times* lamented 'that the first scientific establishment in the empire had obtained a Prince and missed a Philosopher, for its President';[72] the *Spectator* expressed 'great satisfaction' and pointed out the advantage of having as president a man not

eminent in the scientific world since experience had shown, in the case of
Davy and chemistry, that 'an undue leaning is always made to that particular
study which the President himself excels in — to the manifest detriment of all
other branches'.[73]

The first meeting of the Society after the election went well. No threats of
retribution materialized. All seemed to agree that it was 'only against the
system which has hitherto prevailed in the administration of affairs of the
Society and which . . . may still be continued, that the radical members have
any ground for complaint'.[74] The Duke went out of his way to be affable,
announcing that he would

. . . throw open his house for the reception of the Fellows . . . every Wednesday
morning and evening alternately during the session of 1830–31 . . . [and make his
library] accessible to all . . . [members of the Society] for promoting every object
connected with the advancement of science.[75]

Disappointed as Babbage and his cohorts were, they recognized that for the
time being the leadership of the Society was settled and that any press for the
improvements they sought must be made on other fronts. Fitton,[76] in one last
gesture, summed up the case for the Herschel party in a pamphlet[77] outlining
the course of events, refuting various claims made by the ducal party, and
printing a good deal of the correspondence, extracts from minutes, and press
comments on the affair. This 47-page statement was quickly distributed
through the scientific world. Babbage, for example, dispatched a copy to
Professor Benjamin Silliman of Yale, publisher of the influential *American
Journal of Science and the Arts*.[78]

Babbage and H. T. Colebrooke resigned their memberships in the Royal
Society Club,[79] preferring not to continue, with the new hierarchy, the close
social association that dining clubs imposed. Henceforth Charles Babbage
took little interest in the Society or its management. He and other reformers
set about new tasks — personal, scientific, organizational. Some of the group
were content to let time and inevitability bring about the alterations they had
proposed for the Royal Society. Others, more impatient for change, turned
their attention to different means of achieving these ends. One group moved
quickly to organize a second scientific body open to all philosophers. The
British Association for the Advancement of Science was formed in 1831 not
to supplant the Royal Society but to do for science what the more ancient
society seemed unwilling to do. Only five of the signers of the *Times* advertisement — David Brewster, Thomas Brisbane, G. B. Greenough, William
Henry and R. I. Murchison — were present at the first meeting of the new
group but within three years ten others would be among the leaders of the
new association.

Within the Royal Society itself, pressure for change did not cease but no
further challenges for the presidency were made. The old system of carrying
on disputes in private, while presenting a united face to the public, prevailed.
Gradually in time the reforms did come about. Within two years of the 1831

election Sussex was the only non-scientist on the Council. Gilbert's battle to maintain traditional ways of patronage and gentlemanly amateurism was the last one fought for these convictions. In the face of a changing society and of the transformation of natural philosophy into professional science, the Royal Society itself moved gradually from its mixture of friends of science and practitioners of science to one of outright professionals.

Mary Somerville managed to fit into both worlds but her sympathies from the first lay with the professionals.

2. Creation and Publication

The year 1830 marked not only a milestone in the cultivation of British science but a point at which Mary Somerville's role in the enterprise began to take on a new dimension. During the previous decade she had become increasingly part of the London scientific scene, fashioning for herself a unique place in it. In the coming decade this place would be elevated and enhanced. In 1820 she had been a scientific novice. In 1830 she was already beginning to take on some of the aura of a scientific elder, largely because a new generation of scientific men was gathering around her.

Changes in the Somervilles' first scientific circle were inevitable; time and the steady expansion of their scientific acquaintance, not discord, brought these alterations. By 1830 the earlier group had gradually been replaced by one equally distinguished. As with the first, the second set of intimates coupled a good deal of social activity with their scientific communion. Two of Mrs. Somerville's chief mentors among the older generation of London philosophers, Wollaston and Young, were gone before 1830. Their loss was grievous to her but not irreparable; she would never be without superior scientific guides. Several of the first circle — William Blake, Sir John Sebright, Francis Chantrey and Henry Warburton — had by 1830 largely given up scientific interests for other activities but maintained a close social connection with the Somervilles. Warburton and Henry Kater were still active in the Royal Society, Kater retaining his place on the Council in the 1830 election. The junior members of the Somervilles' circle in 1830 — John Herschel, Charles Babbage and William Whewell (1794–1866; F.R.S., 1820) — were by 1830 leading figures in English science and in the Somervilles' second circle. Several relative newcomers to London science were also fast becoming part of the select company that regarded Mrs. Somerville as one of themselves. Charles Lyell (1797–1875; F.R.S., 1826) and Roderick Murchison (1792–1871; F.R.S., 1826) were strongly attached to her and brought her in touch with other geologists and with up-to-date theories about the earth. William Buckland (1784–1856; F.R.S., 1818) and Adam Sedgwick (1785–1873; F.R.S., 1830), frequently in London, thought of her as a friend and scientific colleague. A group of younger men with mathematical and scientific bents, several of them Cambridge friends of her son and including John William Lubbock (1803–1865; F.R.S., 1829), John Elliot Drinkwater (later

J.E.D. Bethune, 1801–1855) and John George Shaw-Lefevre (1797–1879; F.R.S., 1820), were often at the Somervilles' for scientific conversation and company.

Just as her own elders and contemporaries had helped her with advice, information, demonstrations and introductions, Mary Somerville now helped these juniors. And even as her own elders and contemporaries had aided and encouraged her, this new group would do so. The Somervilles' circle in 1830 included three generations of distinguished scientific practitioners and distinguished friends of science. Furthermore, an easy intimacy had existed for over a decade between the Somervilles and the principal philosophers of London, Edinburgh, the provinces, the British universities, Paris, Geneva and scattered foreign centres.

Of this group John Herschel was generally considered by his peers the pre-eminent man of science in England. It was to him that the Somervilles turned for advice in 1830 after Mrs. Somerville completed her Laplace rendition. She had worked at the manuscript for three years under conditions that were far from ideal, as she described them in her memoirs:

> I rose early and made such arrangements with regard to my children and family affairs that I had time to write afterwards not however without many interruptions. A man can always command his time under the plea of business, a woman has no such excuse. At Chelsea I was always supposed to be at home, and as my friends and acquaintances came so much out of their way on purpose to see me, it would have been unkind and ungracious not to receive them. Nevertheless I was sometimes annoyed when in the midst of a difficult problem some one would enter and say I have come to spend a few hours with you. However I learnt by habit to leave a subject and resume it again at once like putting a mark into a book I might be reading, this was the more necessary as there was no fireplace in my little room and I had to write in the drawing room in winter. Frequently I hid my papers as soon as the bell announced a visit lest anyone should discover my secret.[80]

Nor were these distractions her sole hindrances. Letters and papers in the Somerville Collection relating to this first work reveal not only the help she had from willing colleagues but often the sad state of British mathematical resources and scholarship. One of the earliest of these communications is a letter from Augustus De Morgan to Dr. Somerville. De Morgan (1806–1871) had been at Trinity College, Cambridge, with Woronzow Greig and from 1828 to 1831 was Professor of Mathematics at the new University of London. The contents of the letter — from 'University' and dated only 'Tuesday' — suggest that it was written early in his incumbency:

> I send the whole mass of Bailli's writing on the History of Astronomy[81] which I shall be obliged by your presenting to Mrs. Somerville with my respects and informing her that if either of them should be wanted here (the most unlikely thing in the world) I will inform her; that otherwise she is welcome to retain them as long as she pleases.[82]

Charles Babbage, even in the midst of his 1830 battle in the Royal Society, was extremely helpful to her. At her request he lent her a copy — the only one he possessed[83] — of the examples of Lacroix which he, Herschel and George Peacock (1791–1858; F.R.S. 1818) had executed a decade earlier. Further he advised her that De Morgan's 'paper in the Encycl. Metropolitana [on calculus] . . . [was] the best thing on the subject'.[84] At the time difficulties often arose in obtaining copies of needed mathematical works. The Somerville Collection, for example, contains four of Babbage's early mathematical papers — three of which were printed in the *Philosophical Transactions*[85] and one in the *Transactions of the Cambridge Philosophical Society*[86] — not as offprints but written out in clerkly copperplate.[87] From the pertinence of the material, from the occasional pencilled note in Mary Somerville's hand, and from Babbage's approving comments on her mathematics,[88] it is evident that these sheets were consulted during the preparation of her Laplace rendition and carefully preserved afterward.

Her consultants were not all English. On 27 November, almost three years after Laplace's death, she wrote his assistant and calculator, Alexis Bouvard (1767–1843; F.R.S. 1826) at the Paris Observatory, requesting help. She explained that she had been

. . . occupied for a considerable time in writing an account of the works of your illustrious friend . . . which I am advised by some friends to publish. However conscious I am of my inability to do justice to such an arduous undertaking, I am anxious to spare no pains to render my work as complete as possible: it is my wish for this purpose to prefix to it a sketch of the life of the author, but in this country I have not the materials to enable me to execute it. To you, therefore who have for some years enjoyed his confidence and seen his labours, I take the liberty of addressing myself for assistance . . . Every circumstance relating to such a person will be read with avidity in this country, and you can perhaps procure for me, or direct me how to procure anything authentic that has been published. . . .[89]

No account of Laplace's life, however, appears in the printed volume.

On the surface the Somerville family routine exhibited few signs of Mrs. Somerville's new occupations during the years the work was underway. In her autobiography she explained:

I was a considerable time employed in writing this book but I by no means gave up society which would neither have suited Somerville nor me. I made morning visits, we dined out, went to evening parties, and occasionally to the theatre but as soon as it was finished I sent the manuscript to Lord Brougham, requesting that it might be thoroughly examined, criticized, and destroyed to promise if a failure. I was very nervous while it was under examination, and was equally surprised and gratified that Sir John Herschel our greatest astronomer should have found so few errors. . . .[90]

Brougham, who had originally set her on the path of mathematical authorship, during 1829 became more and more caught up in politics and in November 1830 was made Lord Chancellor in Earl Grey's government and immediately elevated to the House of Lords. Even a cursory examination of her manuscript when it was finished was enough to convince him both of its superior qualities and its unsuitability — because of its length — for the sixpence series he was supervising. Brougham asked John Herschel to read the work and make suggestions.[91] Henceforward Herschel was Mary Somerville's chief adviser on this project. It was his encouragement that led to the publication of her first book.

The exact date that Herschel first became involved with Mrs. Somerville's Laplace project is unknown, but the earliest Somerville letter in the Herschel papers makes it clear that it was before 31 March 1829. On that day Mary Somerville wrote him at Leamington, where he was honeymooning, thanking him for taking time to look over her work — when his hours should be spent 'with his bride'. She reported that she was continuing with Laplace's second book and counseled him to take his time with her other sheets.[92] A second letter, in October 1829,[93] indicates that sheets, comments and revisions passed frequently between Slough and Chelsea; in this particular instance Buckland was to serve as messenger.

Early in February 1830, Herschel apparently saw the whole manuscript. On 5 February he wrote Mary Somerville that having completed his essay on 'Sound' for the *Encyclopaedia Metropolitana* 'as a companion to the one on Light of which you have a copy, as I mean you to have of its fellow' and having put his stack of accumulated astronomical reductions in order, he could now turn to an examination of her treatise.[94] He was certainly aware by this date of his friend Babbage's intention to publish his criticisms of the Royal Society and his dissatisfaction with the state of English science. Herschel himself added, sometime in February, a long and significant footnote to the piece on 'Sound'. In it he pointed out that information for this up-to-date article had come chiefly from foreign journals. He praised the speed with which these journals made new scientific findings public, deplored the fact that good science done in England was often recognized abroad earlier than at home, where in turn continental discoveries were largely unstudied. 'It is vain to conceal the melancholy truth', he wrote. 'We are fast dropping behind . . . The causes are at once obvious and deep seated. But this is not the place to discuss them'.[95] Herschel's warning became one of the cardinal arguments of the declinists in the following months.

Even as he expressed these concerns, he found in Mary Somerville's manuscript a superior treatise that could be an important first step in remedying English ignorance of French analysis and physical astronomy. Furthermore, it had come into existence under circumstances that testified to some of the shortcomings he and Babbage saw in English science. Its inspiration had come not from the universities, whose backward mathematical practices these philosophers had criticized for the past two decades, nor from the Royal Society, whose institutional indifference to new methods and needed reforms was under attack, but from proponents of a self-help movement who had commissioned the work primarily for the edification of ambitious mechanics. Its creator had not been a university-educated mathematician but a self-taught Scotswoman. The excellence of her performance was evidence that the national genius still lived and that, given sufficient encouragement, it could produce first-rate fruit.

After this first reading of the whole, he urged the author to go on with her project, lamenting only 'that Laplace has not lived to see this illustration of his great work' and marking two sections for revision (one on virtual velocity and the other on 'a metaphysical nicety . . . hardly worth pencilling your beautiful manuscript for'). He voiced a wishful fear that she might possibly 'give too strong a stimulus to the study of abstract science by this perform-ance'.[96] His letter two weeks later went into more detail about her treatment of virtual velocity, advised her to 'explode' d'Alembert's principle, and urged that she 'double the space' she had given the first chapter, her own prelimi-nary essay on physical astronomy, so essential as background material for 'a beginner'.[97] It is evident from Herschel's letters that her own mathematical competence was never questioned. Her training in the French mathematical tradition was more than sufficient to cope with Laplace's work.

Her principal problem was the insularity of English mathematics. Most of her readers would know only the Newtonian approach. In order for them to understand Laplace in English, they required some simple introduction to the new French analysis and a full but simplified treatment of the materials and methods of French physical astronomy. These Mary Somerville supplied in her 'Preliminary Dissertation'. In the remainder of her work she was deter-mined not to sacrifice any of the essential Laplace.

Three previous attempts had been made to bring Laplace to English readers. A two-volume translation of the French astronomer's more popular *Système du Monde* by John Pond (1767–1836; F.R.S. 1802) appeared in 1809.[98] In 1814 the Rev. John Toplis, a Cambridge graduate and Nottingham schoolmaster, published an English translation and his own 'elucidation' of the first book of the *Mécanique céleste*.[99] Seven years later John Murray brought out Thomas Young's anonymous translation of the same first book.[100]

Mary Somerville's work was far more advanced than either of these two presentations. It covered the first four books of Laplace's *Mécanique cé-leste* and had, in addition, her own long 'Preliminary Dissertation'. She did not limit this introductory 'account . . . to a detail of results but rather . . . endeavour[ed] to explain the methods by which these results are deduced from one general equation of the motion of matter'.[101] Hers is a rendition of Laplace in English, not a mere translation. She employed the Continental notation, included all steps in mathematical demonstrations, carefully defined symbols, compared her approaches with Newtonian ones to the same material, and frequently added investigative findings. Her presentation is in clear, simple English words arranged to produce lucid explanation rather than mathematical muddle. Background material interesting even to mathemati-cally sophisticated readers is included. When pertinent Mrs. Somerville made use of the work of Pontécoulant, Poisson, Delambre and Bouvard, but in general she followed Laplace's presentation closely.

With Brougham's decision not to include Mrs. Somerville's treatise in his library of the Society for the Diffusion of Useful Knowledge, its publication was in jeopardy. Herschel — and no doubt Babbage — must have been anxious lest it be lost to English readers. More than mere personal affection and respect for its author was involved. They realized that the work was one of great potential value to national science and to national pride. The fact that a woman, encouraged and guided through the years by their own reform circle of practising scientists, had produced such a work could be taken as further evidence of the validity of their argument that proper inducement would bring forth significant creation. Not only was Mrs. Somerville's work in itself worthy of their support but the very fact of its existence corroborated many of their assertions about national science and heightened its value in their eyes. To bring about publication of the work would not only serve science but would demonstrate the force of the reformers compared with the indifference — and often ignorance — of the majority of their opponents in the scientific community.

Thus Herschel, always a discriminating judge of scientific and mathematical excellence and a generous friend to colleagues, had added reason to press for publication of the work. In pronouncing it 'a book for posterity, and far above the class for whose instruction it had been intended by Mr. Brougham'[102] he was not only expressing his own opinion of its merits but urging its importance as a much needed addition to the scientific and mathematical volumes then available in English. His sincere praise of the work would be decisive in placing the manuscript with a publisher.

William Somerville, on learning of Brougham's dismissal of his wife's work, moved to salvage what he could of her long exertions. She had entered on the project with little confidence of success but had laboured long and earnestly to complete it. He was determined that her efforts should not go for nought. At this point Dr. Somerville's persistence and his sanguine nature were invaluable assets. By early 1830 he was in touch with his fellow-Scot, John Murray the publisher, whom he reported as 'very desirous to see the work'.[103]

His younger brother, Samuel Charters Somerville, and Murray (1773–1842) had been on intimate terms until the former's death in 1823. The William Somervilles had known the Murrays from at least 1819.[104] Hanover Square was but steps away from 50 Albemarle Street, where the publisher had his establishment. From the informal gatherings of notables in the literary, artistic and political world in Murray's parlor had come the impulse for the Athenaeum Club, of which Murray and Somerville had been founding members in 1824. Murray had, as Charles Lyell pointed out in 1828, a singular 'taste for knowing beforehand what people will read'[105] and had won an enviable reputation as the successful London publisher of Byron, Crabbe, Borrow, Jane Austen, Walter Scott and other literary luminaries. He was also responsible for the *Quarterly Review,* since 1809 a worthy Tory competitor of the liberal *Edinburgh Review.* A Somerville cousin, John Gilbert Lockhart (1794–1854), son-in-law of Sir Walter Scott, was its editor.

In addition to poetry and novels, Murray's list contained numerous popular works — ranging from cookery and travel books to little histories for children — as well as some sound scholarly treatises and an increasing number of writings by scientific men. He himself was not a fellow of any of the London scientific associations, but he had served as 'Bookseller to the Admiralty and to the Board of Longitude' and had published Davy's discourses[106] for the Royal Society and his two popular works.[107, 108] Through his many London contacts, Murray was always well aware of current news and gossip and he recognized the growing public appetite for science and scientific writing. Furthermore he had a strong sense of responsibility to the world of scholarship and letters. Throughout his career he willingly printed works which he considered important even though sales might be slight.

When Dr. Somerville approached him with Mary Somerville's manuscript, Murray had just issued Davy's final book, *Consolations in Travel* (February, 1830), which was doing well, and had ready the first volume of Charles Lyell's *Principles of Geology,* which would do even better. Rival houses were bringing out increasing numbers of scientific books, while scientific series, such as Lardner's Cabinet Cyclopaedia, attracted favourable notice with the promises of forthcoming scientific works. The public appetite for science, already large, was growing.

Murray was accustomed to publishing the works of ladies. His first profitable venture had been Mrs. Rundell's *Domestic Cookery* (1808), still the standby of his list. He had brought out several of Jane Austen's novels and all of Mrs. Markham's histories for children. He had published Mrs. Maria Graham Callcott's accounts of her voyages and, in 1829, anonymously *Bertha's Journal,* a work by Francis Beaufort's sister Harriet. The last two ladies were well known to Mary Somerville. Thus John Murray, for a variety of reasons, appeared a likely start in a search for a publisher for a mathematical work by a lady. It required time, however, for him to reach a decision in this matter.

In the weeks immediately following Herschel's examination of the manuscript, a good many letters passed between that philosopher and the Somervilles. They were 'very sensible of his kindness and gratified by it'[109] and their debt to him would grow as he continued to advise and go over any changes Mrs. Somerville made in the treatise. 'I have only to entreat you not to scruple to give me work', she wrote him, 'for it is my ambition to spare no pains to acquit myself of so bold an undertaking without reproach'.[110] His advice, she assured him, would govern whether or not the material went to press.[111]

The long mathematical letter which Herschel sent her on 9 March 1830 began on an ominous note:

I lose not a moment in forwarding to you a work I have just received from the author. It is a translation of the mécanique céleste with a running comment by (apparently) an able hand. How far its appearance at this juncture may influence your views I am of course incapable of judging but it is a *matter of fact* which you cannot be too early in possession of. . . .[112]

The work which Herschel had just received was the first volume of a translation of Laplace[113] by the American Nathaniel Bowditch (1773–1838; F.R.S. 1818).

Mrs. Somerville replied:

Nothing can be kinder than your early communication of the Mec. cel. I have gone through the commentary as far as the time has permitted, and excellent as the notes are, I confess I am not dismayed as I rather wish to state principles clearly, and to arrive at the results by as easy methods as possible, than to enter into all the mathematical details. I daresay you think me very bold, but I do feel inclined to proceed and to get it into the press as soon as possible.[114]

Work on the manuscript — and inspection of it by Herschel — continued without interruption through the summer. A few hours after the birth of his first child, Herschel wrote Mrs. Somerville of the event, placing this news at the end of a long letter that referred in detail to a passage in her manuscript on 'the general equation of the Elliptic Motions' he thought could not be improved.[115] Six other 1830 letters[116] between Herschel and the Somervilles about the manuscript survive in the Herschel papers.

By August John Murray had decided to undertake publication of the work. He announced his decision in a letter to William Somerville, dated 2 August 1830. In it he indicated no hesitancy in undertaking the publication of such a work by an unknown woman author but he did show concern about the possible number of buyers of so abstract and difficult a treatise:

Mr. Herschells opinion of the excellence of Mrs. Somervilles MSS is a better one than any that I could have obtained and I am perfectly satisfied with it — but I can find no one who can give me data upon which I can calculate the demand for such a work — no publisher can assist me in this — as however Mr. Brougham has said that he can ensure the Sale of 1,500 Copies (which I can not refrain from doubting) I will if you please print one Edition, consisting of 1,500 Copies at my own Cost and risque and in case of their selling I will give the author Two Thirds of the Profits — and after the Sale of these 1500 copies her Copyright shall be entirely the Sole property of the Author — to dispose in any way hereafter that may appear best for her advantage.

By this proposal I mean to try the success of the Work at my own expense, merely for the author's future benefit, without occasioning her any risque or expense.[117]

In a postscript he added that Somerville was at 'liberty to communicate this letter to Mr. Brougham — and if, after trying other publishers, you do not obtain a more satisfactory arrangement, you will find me, still ready to fulfill what I have above proposed'.[118] This generous and open attitude characterized the dealings of Murray and of his son — John Murray II, who succeeded his father in 1843 — with Mrs. Somerville and her family over almost half a century.

Present evidence does not indicate whether or not Dr. Somerville 'tried' other publishers. The arrangements with Murray were settled by September and in the next months Mary Somerville was busy with revisions and correction of proof. On 23 March 1831 her husband wrote John Herschel:

Mrs. Somerville finds the most difficult task she has encountered in bestowing a Title on her work — & the want is the only impediment to Mr. Murray publishing it among his 'In the press &c.' She will be very much obliged to you if you will turn it in your mind . . . She had thought of something in the way of "An analytical view of Laplace's System of the Mechanism of the Heavens" but there is an objection to *analytical* its meaning applying either to analysis of his work, or to mathematical analysis — moreover were the objection removed might the word not be thought presuming if applied to a popular sense of the term but she intended it to be algebraical. . . .[119]

Babbage the previous September had confessed that he himself could not think of a suitable name.[120] Not until late July was one finally found, suggested by the barrister, John Elliot Drinkwater. Drinkwater had been fourth wrangler at Trinity College, Cambridge, in 1823 and had since combined, as so many Trinity mathematicians did, the law and London scientific society. The son of old Fifeshire friends, he was one of the younger men who frequently gathered around Mrs. Somerville for scientific conversation and to meet scientific persons. On 29 July 1831 he wrote her, proposing six titles, of which 'Mechanism of the Heavens' was the third. He added that he thought it 'the neatest, & most striking of the set; my single objection to it may not strike you perhaps very forcibly but I should have wished not to join a Greek word *Mechanism* with a German word *Heavens*'.[121] She had no such scruple and her work became the *The Mechanism of the Heavens.*

In the months before the volume appeared, the Somervilles called on a number of friends for advice, information and help with it. The military engineer, Lieutenant Thomas Drummond (1798–1840), for example, had one of his officers make 'the Necessary Computations' Mrs. Somerville requested and volunteered also the assistance of his superior, Captain T. F. Colby (1784–1862; F.R.S. 1820).[122] Babbage was always helpful in checking for mathematical blunders and looking over proofsheets.[123] Basil Hall sent seven large handwritten sheets of criticism, citing page and line of the proof he had seen, urging Mrs. Somerville to give 'the [Preliminary] Dissertation to the public in a small cheap shape', and predicting '. . . my life for it the work will flash into general circulation . . .'.[124] Even before Herschel examined her completed manuscript in early 1830, Mary Somerville had considered printing its first part separately, a plan he also promoted.[125] Even Lord Brougham found time to go over a few of the sheets. In a note to Somerville he declared, 'Altogether it will be a most wonderful work — perhaps next to the Principia. Let me know the day of publication that I may not get the E[dinburgh] R[eview] into a scrape by a premature notice'.[126]

Despite the labour of seeing the book through the press, the Somervilles still found time for a good deal of entertainment. Among their dinner guests on 2 May 1831 were the Chantreys and a young Scotsman, James David Forbes (1809–1868; F.R.S. 1832),[127] making his first visit to the south. Son of Somerville's Edinburgh banker and brilliant pupil of David Brewster, Forbes brought letters of introduction to various Scottish connections in London scientific circles. In turn, through them, he met a large number of English philosophers. In his journal he mentions that Dr. and Mrs. Somerville received him 'with special kindness', as did Murchison and Babbage when he met them.[128] The philosophers mentioned in Forbes' letters of this date and his journal are those with whom the Somervilles were intimate at the time — Herschel, Babbage, Murchison, Lyell, Drinkwater, Snow Harris (1791–1867; F.R.S. 1831) in London, Whewell, Sedgwick (1785–1873; F.R.S., 1821), George Biddell Airy (1801–1892; F.R.S. 1836), Richard Sheepshanks (1794–1855; F.R.S. 1830), George Peacock (1791–1858; F.R.S. 1818) in Cambridge and Buckland in Oxford.

Forbes not only kept a journal of this first visit south but set down in a small notebook brief comments on — and his assessment of — many of the scientific personages he met. He devotes three pages of this notebook to Mary Somerville — the only woman so noticed — and his words not only give the flavour of her appearance and demeanour in the late spring of 1831 but confirm the authenticity of a portrait previously only said to be of her and attributed to John Jackson, R.A. (1778–1831), who died on 1 June 1831.* Jackson had previously painted two portraits of W. H. Wollaston, one of them a gift to Mrs. Somerville. Forbes wrote:

Mrs. Somerville

Below middle size; fair; countenance not particularly expressive of talent except eyes which are piercing. Shortsighted. Manners of the simplest possible. Age I suppose towards 40,† . . . As to Mrs. S's countenance it is remarkable that Jackson the painter found it extremely difficult to take. I saw one picture, giving a very imperfect likeness & a second, half done, promising still worse. In fact tho' not a very striking [blank space] has the more on that account of an indefinable sort of expression.

Her conversation very simple & pleasing. Simplicity not showing itself in abstaining from scientific subjects with which she is so well acquainted, but in being ready to talk on them all when introduced, with the naivity of a child & the utmost apparent unconsciousness of the rarity of such knowledge as she possesses. So that it requires a moments reflection to be aware that one is hearing something very extraordinary from the mouth of a woman. Of no person does she speak with so much admiration & regard as the late Dr. Wollaston whose t[r]ait of observation she particularly instanced as characteristic of him . . . She mentioned his saying that he regretted that he had not applied himself more to one particular subject. She speaks with frankness of Laplace & his great work (of which I saw the 4th vol. lying in the morning on the drawing room table) & her commentary. She kept up a correspondence with him & one of his last letters before his death mentioned that he was again reading with renewed delight the Principia of Newton.

† A proof of her good looks. — She must be greatly more probably above 50 +[129]

In the larger scientific community Mrs. Somerville had by this date achieved such fame that introductions to her were sought by provincial and other visiting scientists. W. D. Conybeare (1787–1852; F.R.S. 1819), for example, returning to Bristol by way of London after a week in Cambridge,

* The portrait now hangs in the Hall of Somerville College, Oxford.

asked his fellow-Oxonian Charles Lyell (1797–1875; F.R.S. 1826) to introduce him.[130] Lyell, now settled in London and devoting all his time to
geology, had been one of the Somerville circle since coming to the capital. His
journal and letters in 1831–32 are particularly rich in references to the
Somervilles, for early in June 1831, while spending an evening at Chelsea, he
renewed an earlier slight acquaintance with the eldest daughter of Leonard
Horner and fell deeply in love with her. Horner, his health impaired by the
bitter strife and factionalism that gripped the new London University, had
resigned his post as first Warden there and was removing his family to Bonn,
where they could live cheaply and comfortably on a small income. Lyell
followed the young lady to Germany, proposed and was accepted.

After his return to London early in August, he was often at Chelsea, eager
to be with friends who knew and loved his fiancee. His journal and letters
during 1831–32 are filled with references to the Somervilles, many of them
deeply reflective. Since he was himself completing the second volume of his
Principles and Murray was also his publisher, Lyell fell easily in the role of an
experienced author advising a novice. During the next months he took great
interest in promoting the success of Mrs. Somerville's new book. In mid-
August he saw the work at Chelsea 'all done save the index' and was
dismayed that 'about 500 pages . . . [were] algebraic . . . and sealed save to
those deeply initiated'. He hoped that the introduction, which he also
glimpsed would be popular enough 'to waft the heavy cargo on through these
unpromising times'.[131]

Public feeling was running high over reform. The Whigs had kept their
promise and introduced reform legislation soon after taking office. The bill
was defeated, parliament dissolved, and the Whigs returned for a second
time, and a second bill introduced. All these events had taken place since
November 1830 and the country was increasingly caught up in the issue.
'Reform! Reform! . . . Nothing else talked of or dreamt of', Gideon Mantell
noted in his journal early in March 1831.[132] Mary Somerville told Lyell on his
return from Germany that, in all the years in London, 'she never knew
politics . . . [to] so embitter people as this Reform Bill'.[133]

Fears of immediate civil outbreaks had already prompted postponement of
a proposed gathering of philosophers in York from its scheduled meeting in
June 1831 to September. David Brewster, like Herschel and Babbage, had
become deeply concerned with the state of British science and anxious to do
something to repair it. In his review for the *Quarterly* of Babbage's *Reflections,* he had singled out as major weaknesses the teaching of science in
English universities (which neither encouraged the discipline nor rewarded its
practitioners with chairs), the unfair British patent laws (which in effect
penalized the inventor) and the poor management of scientific organizations
and institutions by non-scientists.[134] Now he and others, mainly provincials,
were determined to set up their own national scientific society and had
planned the York meeting.

Among the Somervilles' London circle, only the Murchisons were at York
in September 1831[135] when the new British Association for the Advancement

of Science was formed. William Somerville's name was one of the 121 'Friends of Science' to whom the first circular of the meeting, dated 12 July 1831, was dispatched[136] and Mary Somerville would have been welcomed with him had they attended. Over the next few years the majority of their close scientific friends joined the Association and many assumed positions of leadership in it. Mrs. Somerville herself was regularly invited to its annual meetings and always kept abreast of Association activities. In September 1831, however, this first gathering of philosophers in a new undertaking, caused little stir among the group she saw most often, while the expense of travel discouraged all but necessary journeys by her or her family.

Shortly before the meeting, the appearance of an anonymous pamphlet, *On the Alleged Decline of Science in England by a Foreigner*,[137] aroused some interest in the scientific community. Its author was Gerard Moll, whom the Somervilles had met in Utrecht in 1824. For some months as the controversy went on, Moll had been in correspondence with Michael Faraday, a quiet 'anti-declinarian' — to use Brewster's word — and had prepared a refutation of some of Babbage's charges, together with a listing of weaknesses in Continental science. This tract Faraday edited and privately printed in the late summer of 1831; the identity of its author was generally known.

The pamphlet caused a mild flurry, but excitement over the decline issue was on the wane. The founding of a new scientific body at York gave hope to some of the disaffected, while several minor happenings in science during the year reassured, to some degree, the reformers that they were still a recognized force in British science. Commissioning of the eight Bridgewater Treatises — a rewarding bit of patronage that had fallen into the hands of the P.R.S. — was completed by early March 1831, and half these lucrative appointments went to staunch reformers — Buckland, Whewell, Charles Bell and William Prout — while only one went to an 'ultra', P. M. Roget (1779–1869; F.R.S. 1815). Organized science was also attracting more Royal attention: the King and Queen Adelaide received the Council of the Royal Society as a mark of gracious interest.[138] Further, the Council itself in a four-hour meeting in March discussed two measures of reform directed toward tightening the membership.[139] The Duke of Sussex seemed headed for re-election in November 1831.

Before that date, however, the government moved to give the most public massive recognition yet to the scientific sector: the creation of six 'scientific knights'. The idea was Brougham's; he persuaded Lord Grey and the King that such attention was not only long overdue but would have a bracing effect on the image of British science at home and abroad. The occasion of the King's coronation in September provided opportunity for the creation of a number of new title holders, opportunity that the Whig government was quick to seize and Brougham quick to monopolize in the area of science. In some of the negotiations William Somerville became a willing intermediary.

Mary Somerville's book was scheduled to be published on 1 October 1831. In the second week of September she received several early copies. The first went to John Herschel, not only in gratitude but because he had 'kindly . . .

acceded to . . . [her] wish . . . [that he] review it in the Quarterly'.[140] On the morning of 14 September William Somerville 'at a most unfashionable hour' took the second copy to present to Lord Brougham. Later in the day Somerville wrote Herschel of the occasion, stating that the Lord Chancellor had received the work

> . . . in the most friendly manner identifying himself with its success & he said to effect that two points are very important — to have it reviewed in the Ed[inburgh] & Quarterly — he said that Mrs S must get friend Herschel to review it in the Quarterly & that he should himself [put] this day week a notice of publication for the Ed^r Rev & directions to have the Review written for the N° after next by a competent person. He expressed the greatest satisfaction in hearing how . . . [you had agreed] to review it in the Quarterly. . . .[141]

Brougham went on to declare that 'it was his earnest desire to do everything to patronize and honour Science — but that some time & opportunity were requisite'. In the course of the conversation he asked Somerville to enquire whether Herschel 'would like to have the Guelphic Order conferred upon him'.[142] His father, Sir William Herschel, had been made a knight of this Hanover order in 1816 for his services in astronomy. On his return home, the doctor immediately wrote Herschel, putting the question to him directly. The Somerville family left London the following day for a recuperative stay at Bognor. It was from this resort that negotiations were carried on the next fortnight, often with inopportune delays.

The first occurred when Herschel's reply, dated 16 September, was either slow in reaching Bognor or Somerville's transmittal of it was slow in reaching Brougham or both. Herschel, with great regret, declined the offer. He wrote of his gratitude to Brougham and of the attractiveness of an order that his own father had held but cited as factors more influential in his decision the 'late discussions relative to scientific patronage and the tone which has been taken by some of our more eminent savans in writing on the subject'.[143]

Brougham did not know of this refusal when he wrote William Somerville from Windsor Castle on Sunday, 25 September that not having heard from him, he had sent in Herschel's name with the other five nominees the previous Friday (23 September). He added that he hoped 'Mr H will not be displeased to be one of the number. They are all at the head of some branch of science & all *discoverers* — Babbage — Ivory — Brewster &c.'[144]

His original list had been Babbage, Charles Bell, Brewster, Herschel, James Ivory and John Leslie. Bell, Brewster, Ivory and Leslie were Scottish born. Leslie had taught Brougham and Ivory at Edinburgh. All were sympathetic to the Whigs. Three — Babbage, Bell and Brewster — had signed the *Times* advertisement for Herschel. Only Brewster had been at York. All were eminent figures in the scientific world. Brougham's use of the word 'discoverers' in describing them echoed a distinction made in the *Athenaeum* soon after the 1830 Royal Society election. In a strong article on the current state of British science, the writer had differentiated between the 'labour of [scientific] *investigation*', carried on by 'discoverers', and the 'labour of *application*', carried on by 'inventors' and argued that any English deficiency among the first group was more than compensated by the large number of able Englishmen to be found in the second.[145] It was as 'discoverers', however, that the majority of English scientific practitioners hoped to be known.

The scientific knighthoods being offered were in the minor Royal Guelphic Order, founded in 1815 by George IV to recognize Hanoverian troops at Waterloo and since then usually distributed to military and naval officers ineligible for the Order of the Bath.[146] Sir William Herschel, the only previous Guelphic scientific knight, had been born in Hanover and for that reason was named to this Hanoverian order.

To Babbage the insignificance of the order was an obstacle to acceptance, not on grounds of vanity, but because it slighted science. He was convinced that an order should be created solely for 'those who, without any official position, uninfluenced by any duty, had at their own expense and by their own genius greatly extended the limits of human knowledge'.[147] Brougham agreed with him in principle, but saw no chance of immediate implementation. In his conversation with Babbage, he pointed out that the proposed gestures toward science would be absurd if he — Babbage — and Herschel did not accept. But Babbage was unmoved. Brougham then asked that he at least speak favourably of the plan to Herschel.[148]

William Somerville, when he received Brougham's note from Windsor Castle on 27 September, immediately wrote Herschel, asking him to reconsider his refusal.[149] The next day Brougham — who either had not yet heard from Somerville or who had not taken his report of Herschel's reluctance seriously — wrote the doctor from the House of Lords:

I should be extremely sorry after the very kind way in which H.M. received my proposition that Mr. H[erschel] did not accept — He [the King] spoke kindly of all his family. It is one thing not to ask and another thing to refuse when offered and still more when pressed.[150]

When Somerville received this note on 29 September he instantly sent a copy off to Herschel, advising him that he 'had no choice . . . that in accepting the title the Honor would be done to it by you & I really think your acquiescence would tend to promote the good of Science'.[151] The next day Herschel wrote that he would submit. He 'frankly' declared himself 'gratified and encouraged by the proposed honour & more so by the quarter in which the proposal originated and the manner in which it has been offered'.[152] On 4 October he sent Somerville word that he had written Brougham and accepted.[153] Brougham's relief must have been immense. He had already struggled to fill one empty place — Babbage's — and had found not a scientist but a reformer, Harris Nicolas (1788–1848), whose efforts had gone chiefly to the records commission. Herschel was the best known of the six knights but would himself take more pleasure in the baronetcy given him by Queen Victoria in 1838.

Not until 5 November 1831 did preliminary announcements of the *Mechanism of the Heavens* appear in the weeklies.[154, 155] From the first Mary Somerville publicly acknowledged authorship, an unusual step at a time when most women — even the well-regarded Jane Marcet — still preferred the anonymity of 'By a Lady' or 'By the Authoress of ———'. Mrs. Somerville, like her friends Maria Edgeworth and Joanna Baillie, proudly but unostentatiously claimed her work.

And the credit she received was all directed to her alone. Charles Lyell in a letter to his fiancee on 23 August 1831 made several penetrating remarks about this subject. In drawing a comparison between Mrs. Somerville, whom he loved and honoured, and Madame de Staël, whom he looked upon 'as a phenomenon — a giantess almost rivalling any male giant of her age', he wrote:

Had our friend Mrs. Somerville been married to La Place, or some mathematician, we should never have heard of her work. She would have merged it with her husband's, and passed it off as his. Not so De Stael, she would have been a jealous rival . . . A man may desire fame, reputation, and even glory, for the sake of sharing it with one he loves. A woman cannot share it with her husband, it will be the utmost she can do not to make him of less importance by it. . . .[156]

In Mary Somerville's case, William Somerville was never a rival but always a loyal and often efficient assistant. His stature was enhanced, not diminished, by his wife's fame, in which he always took great pride and delight. Throughout their marriage, his energies and his efforts were exerted on her behalf and on behalf of her public recognition and reward. Had he himself been a scientist the picture might have been different.

The *Mechanism of the Heavens* was finally published at the beginning of November 1831. The volume Murray produced was a handsome one, beautifully printed by William Clowes on good paper, and priced at £1. 10s. Its 703 pages encompassed her 'Preliminary Dissertation' (70 pages), a short statement about physical astronomy (3 pages), and Laplace's first four books arranged in 34 chapters of varying lengths. The first book (six chapters, 144 pages) is a treatise on dynamics, as given in the *Mécanique céleste*. The second, and by far longest (14 chapters, 266 pages) considers the effects of universal gravitation on planetary motions and orbits. Book III (six chapters, 90 pages) is concerned with lunar theory, Book IV (nine chapters 109 pages) with satellites. There are 116 figures, a good index (11 pages) and two pages of errata, mostly omissions of numbers and symbols in mathematical formulae. The work was dedicated to Lord Brougham, who had first suggested it, in simple words that avoided flattery and gave a succinct account of his original intention. Mrs. Somerville's closing words in this inscription, dated 21 July 1831, were, 'To concur with that Society in the diffusion of useful knowledge, would be the highest ambition of the Author.'.[157]

3. Reception

On the evening after copies of the *Mechanism of the Heavens* appeared in the bookshops the Somervilles attended a large party at Lansdowne House. Dr. Henry Holland, whom Mrs. Somerville identified in an unprinted passage in her autobiography merely as 'a celebrated physician who often presided at scientific meetings and wrote clever articles in the Reviews', came up to say

It is a pity you published that book you have made a sad mistake with regard to the effect of the air on falling bodies in the very beginning of the book which vitiated all the rest. I am sorry for you.[158]

She adds,

. . . I was thunderstruck and fairly lost my head or I should have seen at once that he was speaking nonsense besides I might have been certain that neither Mr Herschel nor Lord Bougham would have overlooked so gross an error but I was confused and spent a very unhappy evening. This was not the only attack . . . : a Mr. Buller member for someplace I have forgotten in the west of England spoke of . . . my book with sovereign contempt. I was much annoyed more so than I ought to have been for he showed that he was totally ignorant of the state of science.[159]

Notes for her autobiography carry the story farther:

. . . a few days after . . . [publication] a member of the house of Commons rose and said a book has been published by a Mrs. Somerville about which a great deal of nonsense will be spoken as if it was clever &c &c whereupon Mr Henry Warburton who was known to be an excellent mathematician rose and said Mr Speaker, I have read the book (a mistake I fear) & I can only say that there are not more than 5 men in Great Britain capable to have written it. . . .[160]

Holland's and Buller's remarks, and one review, were the only detractive comments directed toward Mrs. Somerville and her work. Its reception was highly laudatory and its author praised in Britain and abroad.

Laplace's *Mécanique céleste* was, in the view of the day, second only to the *Principia* as a work of genius. Mrs. Somerville, in undertaking to bring it to English readers, had set herself a task that few would or could undertake. A failure would be disappointing but understandable. A success would take her to the highest reaches. To her contemporaries Mary Somerville's achievement was not merely the rendition in English of a French book on mathematics; it was the rendition in English of the most important work — next to Newton's — yet produced by the mind of man. To most of the British public her work itself would remain unknown. What they would know was the reputation which it earned for her, a reputation that made them regard her with awe and pride.

A review of the reception of the work in the press, taken together with letters in the Somerville and Herschel papers, not only indicates how her book was received but how — in a transitional period between venal puffery and honest reviewing — books came to public notice. For an abstruse first work by an unknown woman author, the *Mechanism* received considerable attention. In mid-December, soon after publication, the *Literary Gazette* hailed it as 'the readiest means . . . possible' of acquiring an understanding of the mechanism of the heavens, saluted 'the gigantic experiment our country-woman undertook to perform . . . to give the world a succinct, profound, but, at the same time, as popular a view as possible' of Laplace, and praised her 'luminous and precise style'. The reviewer urged that the 'beautiful preliminary dissertation . . . be printed separately, for the delight and in-struction of thousands of readers, young and old. . .'.[161]

Its reception by the rival *Athenaeum* was quite different. Not until 21 Janu-ary did a review appear and then it was mockingly and patronizingly derisive in tone. Its opening paragraphs gave no hint of what was to come since they deplored the neglect of Laplace in England, citing it as evidence of the decline of science. The remainder of the review, however, was given over to ridicule: of Brougham and the Society for the Diffusion of Useful Knowledge for their illusion that the working class wished to be introduced to Laplace and for their foolishness in choosing a woman to undertake such a task; of Murray for 'the splendour of the typography of . . . [a] volume . . . [intended for] the hands of the unwashed'; of Mrs. Somerville for attempting to reduce 'the very spirit and essence of . . . [Laplace's] four quarto volumes and supple-ments . . . [to] a single octavo; and finally and gratuitously, of the 'confirmed blue[s] of the United Kingdom . . . [who would] prattle . . . [of the *Mechan-ism* and] then, when the novelty of its youth has passed away . . . [assign it a conspicuous place in the library where its retirement would be] perfect . . . [and] uninterrupted'.[162] Most of the long piece went to diminishing Mary Somerville's achievements. Its words and tone demonstrate disdain for learned women, while its mathematical comments reveal its writer's inade-quacies in that discipline.

Its author may well have been Charles Buller (1806–1848), M.P. for West Looe in Cornwall, a frequent contributor to the *Athenaeum*. He was notori-ous for his ready taunts and a prankish disposition that sometimes obscured a generous and liberal nature. Why he should — if indeed he did — attack Mrs. Somerville and her book in the press and in the House of Com-mons — for many of the words of the review echo those said to have been spoken in the House — is unknown and must lie in his own personal crochets. Buller was at Trinity College, Cambridge in 1825 and may have known Woronzow Greig. He was one of the earliest of the 'Apostles' and kept up this college association in later life; he was among the six 'Apostles' who subsidized the *Athenaeum* after F. D. Maurice, founder of the society, purchased the weekly in 1828.[163] And in 1837 Buller, a man liberal in politics, spoke harshly against Mrs. Somerville's civil pension.

The *Athenaeum* remained alone in its disparagement of the *Mechanism*. The veteran Whig *Monthly Review* in its notice not only praised the book but went on to endorse female education and to list the intellectual accomplishments of a number of English and foreign ladies.[164] Among practicing scientists the book was received with acclaim. Robert Jameson did not review it in his *Edinburgh New Philosophical Journal,* but its April-October number carried a long extract from Mrs. Somerville's 'admirable work'.[165]

It was in the quarterlies, however that serious literary reputations were made and wide notice taken of philosophical matters. The two greatest of these — the *Edinburgh Review* and the *Quarterly Review* — both gave the work lavish and prominent praise. Lesser reviews — the *Westminster,* the *Eclectic,* the *British,* the *Foreign* and the *Foreign Quarterly* — did not address this specialized work. Herschel's report for the *Quarterly* was already well underway before the publication date. He wrote William Somerville on 16 September that he was setting 'about the review forthwith' and added, 'I think of taking that opportunity to say a few words about Bowditch's work which is too great an undertaking to drop like a dead weight on the public'.[166]

Brougham wrote Macvey Napier (1776–1847; F.R.S. 1817), editor of the *Edinburgh Review,* about the volume before a copy arrived in Edinburgh (early in December 1831). Napier also heard from a 'Dr. Morrison and his son . . . on the subject of an article upon this work for the E. Review' but, as he wrote William Somerville on 5 December, he wished to place the book in the hands of John Leslie or Thomas Galloway (1796–1841; F.R.S. 1834). If Leslie's engagements prevented his having a review ready for the March number, he intended to turn the task over to the younger Scotch mathematician Galloway. Napier added that David Brewster was the only 'other person . . to whom I should chuse to commit the book' but that as Brewster wrote regularly for the *Quarterly* — the implication is that Napier expected him to examine it for them — he felt it would be 'indelicate and improper . . . to apply to him.'[167]

The book finally went to Galloway, a favourite student of Mrs. Somerville's old mentor William Wallace and, since the previous year, Wallace's son-in-law. Galloway first learned French mathematics in 1811 from two French prisoners-of-war held near his home, continued the study at Edinburgh, and since 1823 had taught at Sandhurst. He had written the treatise on astronomy for the new seventh edition of the *Encyclopaedia Britannica* and a recent piece on the subject for the *Edinburgh Encyclopaedia.* Napier mentioned both to Dr. Somerville should he wish to 'see specimens of his [Galloway's] talents for scientific writing', assuring his old friend — both he and Somerville had become fellows of the Royal Society in 1817 — at the same time that he, as editor, would 'of course state to him on what way I wish the article written'.[168]

Galloway's review[169] was the first article in the April number of the *Edinburgh.* In it he gave a cogent summary of Laplace's *Mécanique céleste* and of Mrs. Somerville's rendition, including her additions and omissions. Each chapter in the *Mechanism* was briefly summarized, the whole was lauded in grave and measured terms. Special praise was accorded Mary Somerville herself; the review begins:

This unquestionably is one of the most remarkable works that female intellect ever produced, in any age or country; and with respect to the present day, we hazard little in saying that Mrs. Somerville is the only individual of her sex in the world who could have written it.[170]

It ends on the same note of personal praise for the author.

The effect of this favourable reception was described at the time by Charles Lyell in a journal entry that unconsciously hints at a declinist position:

. . . Such inquiries for her [Mary Somerville] by strangers, as since her work was reviewed in the Edinburgh (for Herschel's in Q[uarterly] R[eview] is not yet out) she has been the great lion in town. How strange that people never knew before that she could have done all this ten years sooner. They knew it in France.[171]

This comment was more cheerful than the one he had made late in November, shortly after the book came out. Then he wrote that she was looking ill and anxious — her son had been suddenly and seriously unwell in Edinburgh but was recovering and she had found little cure at Bognor — and added, 'I think the book has done her no good, and when I see it, I am sure that such a work must have been too great an effort for any one in the time. It was a gigantic undertaking'.[172]

Herschel's account did not appear until the July number of the *Quarterly Review*[173] and as expected coupled Bowditch's translation with Mrs. Somerville's rendition of Laplace. To the former, however, he devoted only the final long paragraph, taking the opportunity in it to call attention to 'the liberal offer of the American Academy of Arts and Sciences, to print the whole at their expense' and declaring the work 'even in its present incomplete state . . . [to be] highly creditable to American science'.[174] Herschel not only ended his review, which was highly laudatory of Mrs. Somerville, with this mild reminder of some of the niggardliness of British institutions toward science but began it with a gentle but implacable comparison of developments, since the time of Newton, in French, English, and Scottish mathematics. He devoted the first six of eleven full pages to a statement of some of the post-Newtonian demands on mathematics and of the 'triumphs of both pure and applied mathematics abroad . . . [and] their decline, and indeed, all but total extinction at home'.[175] He considered the difficulties of those British mathematicians — Playfair, Ivory, Babbage, Airy, Lubbock, and Challis — who had moved to French analysis and pointed out the value of Mrs. Somerville's work. This part of his review comes as near being a public statement of his declinist views as he ever made.

Not until the mid-point of his review did he refer to Mary Somerville's sex and then only to mention her previous experimental work and to declare that nothing 'beyond the name in the title-page . . . [could be found in the work] to remind us of its coming from a female hand'.[176] He praised her whole-hearted and disinterested commitment to science and regretted only her omission of some of Gauss's work and a theorem of Lambert and her 'habitual laxity of language evidently originating in so complete a familiarity with the *quantities* concerned, as to induce a disregard for the *words* by which they are designated.[177] His impression of the whole, he declared to be one of unfeigned delight and . . . astonishment'.[178]

The responses of friends were equally warming. Almost six dozen presentation copies went to various institutions and persons from the author. Several letters thanking her for the gift mentioned the recent charges of a decline in English science. William Wallace wrote that the book should 'incite' him to exertion and added, 'I was never an absolute believer in your friend Mr. Babbage's Theory that Science is on the Decline in Britain. Your book certainly gives it no support'.[179]

Francis Baily expressed these sentiment when he called the work 'highly honourable to herself . . . of great benefit to the public; not only as tending to their improvement, but removing the imputation of the DECLINE OF SCIENCE in this country'.[180] Those who knew her less well were apt to dwell on their astonishment that one of her sex should have produced such a work. 'I confess.' James Ivory wrote in thanks for his copy, 'I was somewhat astonished to receive a Book treating of so many difficult and abstruse subjects, written by a Lady with so much clearness and method'.[181]

In Paris the book was received with delight. Biot immediately prepared a long and admiring account of it which was published in the January issue of the *Journal des savans*.[182] Writing to Mrs. Somerville of this piece in May 1832 he gave a rollicking description of the reception of his report on the book at the Academy *séance* of 13 February 1832.[183]

. . . Le plus amusant pour moi de cette rencontre, c'était de voir nos plus graves confrères, par example, Lacroix et Legendre, qui certes ne sont pas des esprits légers, ni galants d'habitude, ni faciles à émouvoir, me gourmander, comme ils faisaient à chaque séance, de ce que je tardais tant à faire mon rapport, de ce que j'y mettais tant d'insouciance et si peu de grâce; enfin, Madame, c'était une conquête intellectuelle complète. . . .[184]

Two other Parisian scientists sent their thanks in the spring and summer. Poisson praised the work and urged her to continue through Laplace's other books;[185] he also included a copy of his new work on the theory of capillary action[186] for her in a packet on its way to the Royal Society. Gay-Lussac sent his thanks in a letter introducing a colleague, Viscount Hericart de Thury (1776–1854), in London to visit various establishments and hoping to see Chelsea Hospital. Of Mrs. Somerville's book he wrote, 'Quoiqu'il soit beaucoup trop savant pour moi, je n'y attache pas moins un très grand prix; c'est pour moi un trop aimable souvenir'.[187]

As soon as the book appeared Mrs. Somerville 'begged . . . [Charles Lyell] to send a copy . . . to Bonn', which he as Secretary of the Geological Society included in a 'parcel to [the geologist] Von Oeynhausen'.[188] Soon after its arrival there, Leonard Horner lent the copy to young Professor Julius Plücker (1801–1868; F.R.S. 1831), who after reading the 'Preliminary Dissertation' carefully and the remainder of the book more rapidly, volunteered to undertake a review of it for one of the Continental 'journaux littéraires les plus éstimés'.[189]

From Cambridge Whewell sent her a copy of his original sonnet, 'To Mrs. Somerville, on her Mechanism of the Heavens';[190] she had presented a copy of her book to him and another to Trinity College Library. A third went to the Cambridge Philosophical Society, of which Adam Sedgwick was president. Sedgwick confided to Charles Lyell that he had 'proposed to the . . . Society . . . to elect her, by acclamation, member on receipt of the book but some objected to the manner of doing it'.[191]

Lyell took a great practical interest in the book. His own second volume of the *Principles of Geology* was selling well but as early as mid-November 1831 he was concerned about the stagnation in trade, along with the depressing effects of the general unrest over the issue of reform and the spread of the cholera that had reached northeast England in October 1831. Lockhart predicted at the time that another six months of such conditions 'would make every bookseller bankrupt, except Longman and Murray'.[192] A conversation with Adam Sedgwick some weeks later did little to cheer Lyell. Sedgwick declared Mrs. Somerville's book to be 'most decidedly the most remarkable work published by any woman since the revival of learning' but went on to say that 'few men at Cambridge can go far enough to enter into it'.[193]

Sales had not improved by mid-February when Lyell reported that young John Murray had said that

. . . Mrs. Somerville's book does not sell at all, and that Brougham had made great professions, both to him and his father, of what he would do to push it, whereas he has not taken the slightest pains to help it. The surprising and fine part of the work is, that everything which she there shows she so perfectly knows, was acquired for mere pleasure, just as others read a poem. The State might award her 5,000 l. for the benefit conferred by a woman who could thus teach what Johnson justly called "the most overbearing of all aristocracies, that of mathematicians," how most of them can be equalled and surpassed by a lady who was merely reading for her amusement. It is true that there is not much display of *inventive* power in the book — not that kind of power, for example, which enabled Babbage to invent his machine, but there is some I understand.[194]

He feared that Mrs. Somerville's new volume, the *Preliminary Dissertation to the 'Mechanism of the Heavens'* [195] would sell no better than the *Mechanism* itself.[196] She had long considered bringing out this first part separately and by the first week in March 1832 had it ready for the press. William Somerville at the time described it as 'considerably enlarged, mid-popular & I hope it will have a good sale — as the only chance of her deriving any advantage from her toils',[197] but the book as printed by Clowes was no more than a reprint of the original dissertation. Murray took no part in the transaction. Among the Somerville papers there is no evidence of the number of copies printed, costs, sales or profits. From letters in the Somerville Collection it is clear that at least a few copies went as gifts.[198] Almost immediately after this section came out, Carey and Lea in Philadelphia pirated the work, issuing the same text in smaller format.[199] No American edition of the *Mechanism* was published.

On Valentine's Day 1832 George Peacock wrote Mary Somerville from Cambridge that he and Whewell had made the *Mechanism* a classbook for their advanced mathematics students. 'I consider this', Mrs. Somerville writes in her memoirs, 'as the highest honour I ever received, at the time I was no less sensible of it, and was most grateful'.[200] Peacock predicted that the book would 'immediately become an essential work to those of our students who aspire to the highest places in our examinations'.[201] Lyell rejoiced because Cambridge approval guaranteed steady, if slow, sales and because the book would be useful.[202]

The criticism which he mentions as hearing — that Mrs. Somerville had done nothing in the way of invention but was merely 'an expounder of Laplace'[203] — would recur throughout her career. She herself was convinced that women, as a sex, had no creative intellectual powers. In the second manuscript of her memoirs she wrote:

In the climax of my great success, the approbation of some of the first scientific men of the age and of the public in general I was highly gratified, but much less elated than might have been expected, for although I had recorded in a clear point of view some of the most refined and difficult analytical processes and astronomical discoveries, I was conscious that I had made no discovery myself, that I had no originality. I have perseverance and intelligence but no genius, that spark from heaven is not granted to the sex, we are of the earth, earthy, whether higher powers may be allotted to us in another state of existence God knows, original genius in science at least is hopeless in this.[204]

Creative or not, she was convinced that this work would be her most enduring. 'All my other books will be soon forgotten', she wrote in her ninetieth year, 'by this my name will be alone remembered'.[205]   ·

The Royal Society, in its new mood of change, moved quickly to recognize her achievement. She presented them with a copy of her work and they sent proper thanks.[206] Then, as early as 13 February Lyell reported that there was 'a little talk of getting a bust from Chantrey of Mrs. Somerville, to be put up in the Royal Society at their expense'.[207] Davies Gilbert approached Dr. Somerville on the matter shortly afterward at the first of the 1831–32 *soirées* given for the Fellows by the P.R.S. at his apartments in Kensington Palace. He asked if Mrs. Somerville might object 'to a proposal that he & the Duke of Sussex were determined to carry into effect — with her concurrence — namely to allow her bust (to be made by Chantrey) to be placed in the R. Soc.'[208]

By 18 February all was settled and on the following day the Secretary, J. G. Children, wrote Somerville a formal letter stating that the Duke of Sussex perfectly approved the proposal and had directed Children to communicate with Chantrey.[209] Furthermore, the Duke, with a handsome subscription of 21 guineas, placed his name at the head of the list of contributing admirers. Sixty-four subscribers pledged £156.10 of the £200[210] the sculptor usually charged;[211] Chantrey declared that he would, out of admiration, make up any shortage himself. About half the subscribers can be characterized as 'reformers' in terms of differences in the Society, slightly fewer as 'neutrals' and only eight as 'ultras'.

The Geological Society took no official notice of a work so remote from their specialist interests but they they did elect William Somerville a member of the Council in February 1832; he served for two years. He had done no work in geology at all but had an amateur's interest in hearing about the subject. He and his wife were now on intimate terms with leading geologists of the day and she had readily adopted the new views of the subject they proposed.[212] Through her husband they saluted her.

Not only was Mrs. Somerville's work lauded by learned gentlemen but it received the praise of many ladies, despite their protestations that it was far above their understanding. Miss Mary Berry, Joanna Baillie, Maria Edgeworth, Lady Callcott, Mrs. Henry Kater and the recently-elevated Lady Herschel sent their congratulations. Lady Herschel expressed the sentiments of many of them:

> You must allow us [to] congratulate *you all* on the splendid & almost *everlasting* mark of esteem which the Royal Society has conferred for the first time on a female & which you may have the most *honest consciousness* before God & Man, of having richly deserved — I propose that we poor women, whom you have left so far in the background, shall raise a monument also, to shew our Sincere love for you — & one ready confession that you have not abjured your sex, while soaring far above it. . . .[213]

From Tunbridge Wells, where the ailing Katers had gone for their health, an even older friend, Mary Frances Kater, wrote:

> I cannot tell you the delight with which we heard . . . of the intention of placing your bust in the meeting room of the Royal Society — They only do themselves honor by conferring on you the only one they had it in their power to bestow — I am afraid of troubling your modesty (that *"rare thing in women"* as you know Sir James South so gallantly said) so I will not tell you all we have said & thought about it, but I must tell you Edward's [the Katers' son] exclamation "Now then her daughters may look down upon Dukes!"[214]

These letters reflect the feelings about Mary Somerville held by the wives, daughters and female relatives of many of the scientific men who themselves esteemed her. Bluestockings might be deplored, derided and condemned, but Mrs. Somerville had only their love and approbation. Her scientific and mathematical talents appeared unique, a gift of the Creator, as great beauty in a woman is unique. Her studies seemed wholly suited to these gifts. Honours to her were honours to the sex. No hint of jealousy or alienation disturbs the affectionate regard their words convey. Although they felt Mrs. Somerville's intellectual powers far above their own, they sought without constraint her advice on ordinary matters and continued, despite her growing fame, to seek and enjoy her company. She earned their good opinion through her sweetness of character, her quick sympathy, her evident feminity, and her readiness to take part in and to relish the same social and domestic activities that filled their days. Neither she nor they condescended to each other but were joined in cordial and unaffected bonds.

Her close Cambridge admirers — Whewell, Peacock, Airy, Sheepshanks and Sedgwick — having failed to persuade the Cambridge Philosophical Society to make her an honorary member, were still determined that there should be some Cambridge homage paid to her achievement in creating the Laplace rendition. Of all British universities, Cambridge had the proudest mathematical tradition. It was there that the first calls for mathematical reform had sounded in the second decade of the century. Cambridge men had been leaders in such innovations as the Geological and Astronomical societies. They were strongly represented in the reform element in the Royal Society. Whewell, Sedgwick and Sheepshanks had signed the advertisement for Herschel's election, Peacock and Airy had been sympathetic to several of the changes advocated in national science. The university itself was deeply conservative but its scientific members tended to be liberal, sometimes radical.

Always hospitable, the Cambridge group determined to invite the Somervilles to be their guests in that university town so that they might receive the same academical entertainment proffered other visiting savants. An occasion of social pleasure would also become, in the eyes of the world, a public celebration of Mary Somerville's praiseworthy achievement, a public acknowledgement of their admiration for and endorsement of her work. Airy spoke to Somerville of a possible visit when he was in London in March and suggested that they come before 13 April (the beginning of vacation) or after 2 May (its end) and be the Airys' guests at the new Observatory.[215] But that site was more than a mile from the University and Sedgwick insisted that they stay instead in the absent Sheepshanks's rooms at Trinity.[216] This plan was adopted[217] and the week of 9 April set for the visit.

It was one of utter delight for guests and hosts. Mary Somerville was awed and pleased at the honour shown her and at the opportunity to stay in a spot so filled with great mathematical memories, one also in which her son had spent pleasant years. She was received as a celebrity. Young William Rowan Hamilton (1805–1865), by chance in Cambridge at the same time, was mightily impressed at meeting her.[218] The visit is well documented, and a day-by-day account can be compiled from various sources. In letters to Cambridge friends she afterward referred to '*our* College' and the happy week spent there.

When the Somervilles returned to London William Somerville's name was again placed in the List of Managers of the Royal Institution,[219] another indirect gesture of recognition to his wife. Mrs. Somerville's opinion was now sought on a variety of policy matters in science, and in April and May, she appears from outside evidence — there is none in her papers — to have been consulted about plans for the second meeting of the new British Association. It was scheduled for Oxford in June. William Buckland was to act as President, Charles Daubeny was in charge of local arrangements. In the spring Buckland wrote Murchison, who was serving on the Committee of the Association:

> I was most anxious to see you to talk over the proposed meeting of the British Association at Oxford in June. Everybody whom I spoke to on the subject agreed that, if the meeting is to be of scientific utility, ladies ought not to attend the reading of the papers — especially in a place like Oxford — as it would at once turn the thing into a sort of Albemarle-dilettanti-meeting [refering to the lectures at the Royal Institution], instead of a serious philosophical union of working men. I did not see Mrs. Somerville; but her husband decidedly led me to infer that such is her opinion of the matter, and he further fears she will not come at all.[220]

Ladies, Mrs. Murchison among them, had been at York and had served notice of intention to be at Oxford. The success of Mrs. Somerville's book, the wide acclaim being given it in the highest spheres of science had altered the place women now held in the scientific world. It was imperative that a new organization, struggling to make a place for itself, should know her views on the matter. Babbage, who had not been at York, intended to go to Oxford and had no hesitancy in sending Daubeny a few 'hints' about arranging the meeting. The second of these dealt with the admission of ladies:

> I think also that *ladies* ought to be admitted at some kind of assembly; remember the dark eyes and fair faces you saw at York and pray remember that we absent philosophers sigh over the eloquent descriptions we have heard of their enchanting smiles. It is of more importance than perhaps you may imagine to enlist the ladies in our cause and the male residents throughout the country will attend in greater numbers if their wives and daughters can take some share of the pleasure. If you will only set up an evening conversation for them at OXFORD I will try to start a ball for them at Cambridge.[221]

The passion and fashion for science which pervaded the upper classes and the attention given Mrs. Somerville and her book seem to have fused in the late spring into a concerted demand by ladies — one supported by many encouraging males in scientific circles — for new opportunities to take part in scientific affairs. They had long been part of audiences at the Royal Institution and at performances by itinerant scientific lectures, had been allowed an interest in the Horticultural Society since its founding in 1804 and had been admitted as Fellows in that Society for some years. Their acceptance at York and their pending acceptance at Oxford — for they were not forbidden to come and shortly the problem was solved for a time by the issuance of 'Ladies tickets' — led to intensified female impulse toward science. Mrs. Somerville's extraordinary feat in producing a successful mathematical work of high seriousness and the unique acclaim given her by the Fellows of the Royal

Society in commissioning the Chantrey bust inspired both men and women to go beyond the merely fashionable in science.

The next step came in the spring of 1832 in the very fortnight that defeat of the Reform Bill on 7 May gave rise to a momentous national crisis and outcry that led finally, on 4 June, to passage of the Bill and major reform of the franchise. A much smaller reform — yet one important to women — took place at the recently opened King's College of the new University of London. Charles Lyell, named its first professor of geology in 1831, had scheduled his inaugural lectures for May 1832. In January of that year he published the second volume of his *Principles of Geology*. Its great success not only heightened interest in his coming lectures but persuaded him to give up his chair as soon as possible to devote his full time to writing and geologizing. As word of his intention spread in the scientific community, the size of his potential lecture audience grew. A number of ladies appealed directly to an officer of King's College for admission to the talks. He consulted Lyell, who — surprisingly in the light of his many pleasant scientific associations with women but understandable in view of his uncertainty as a novice professor — rejected the notion on the grounds that women in a classroom would be 'unacademical'.[222] Nonetheless a few ladies turned up among the eighty persons at his first lecture on 1 May, a day of heavy rain and consequently poor attendance.

The following day when the Geological Society met, there were, as he noted in his journal, 'grand disputes . . . about the propriety of admitting ladies to my lectures'.[223] Babbage and Murchison were especially insistent on the point. Two days later, when Lyell delivered a hastily revised second lecture, almost 300 persons — both men and women — were present. Among them were Mrs. Murchison, Mrs. Somerville and the two Somerville daughters.[224] Later that day Mary Somerville wrote Whewell in Cambridge, 'It is decided by the Council of the University that ladies are to be admitted to the whole course, so you can see what in[va]sions we are making on the laws of learned societies, reform is nothing to it'.[225]

Lyell, quickly amenable to the change, immediately urged the Somervilles and several other ladies to continue to attend,[226] an invitation they readily accepted. Mary Somerville declared herself 'charmed with the effect' the lectures were having on her daughters and told Lyell that whereas they 'would never read before . . . they [had] set to work in earnest' on his two volumes.[227] The girls attended faithfully, although seventeen-year-old Martha confessed to her brother a year later that she thought Lyell 'a most tiresome lecturer — at least I was never so tired of anything in my life, as those in the King's College'.[228] Before the end of the series in June 1832, traditionalists had induced the Council of King's College — in a hasty step that Lyell deplored — to forbid future entrance of women to lectures lest the attention of young students, all male, be diverted.[229] Nevertheless, the admission of women to Lyell's lectures marks an important first step in the gradual opening of universities to female students.

Mary Somerville always supported in word and often in deed the struggle for wider opportunities for women. She did not, however, go to Oxford in June 1832 for the B.A.A.S. meeting. She was ill in late May, tired from the long labour of her book and the excitement of its success. She may have been influenced also by John Herschel, who stood aloof from the new organization until its meeting the next year in Cambridge. A far more probable reason for her absence was the Somervilles' chronic shortage of money. Jaunts, even one of only a week, were costly for ladies: transportation of the travelling party and one or two trunks; upkeep during their stay; an appearance in accord with her position in life. Dr. Somerville or her son and a lady's maid would have to accompany her. On 12 June she wrote her old intimate Mrs. Leonard Horner, struggling in Bonn to rear a large family of daughters on a small income, that, 'We remain at home indeed we shall be stationary all summer moving is so expensive'.[230] Nothing in Mary Somerville's character and none of her other actions lend validity to the statement of Buckland's daughter half a century later in her biography of her distinguished father: 'In the end Mrs. Somerville decided not to attend the meeting [of the B.A.A.S. in Oxford] for fear that her presence should encourage less capable representatives of her sex to be present'.[231]

CHAPTER 6

THE SECOND STAY ABROAD

1. Paris, 1832

In the summer of 1832 when cholera was at its height in London,[1] Mary Somerville suffered 'two severe bilious . . . attacks at Chelsea'[2] from which her recovery was slow and that left her thin and weak. Her husband insisted that she have a complete change of air and scene and they fixed upon a jaunt to Paris. Mrs. Somerville would have opportunity there for rest and diversion in surroundings she already knew. She, whose Laplace rendition had won the approval of Parisian savants, could expect a cordial reception, interesting invitations, and helpful conversations. Their daughters, now 15 and 17, had never been abroad and would profit from exposure to the culture and style of the French capital. Both girls spoke and wrote French and German well enough to serve on occasion as translators for their mother's scientific friends[3] and were, for their years, remarkably easy in intellectual society. They had no flair for mathematics or science, but could converse about literature and art. Neither girl was pretty — Martha resembled her father, Mary was more like her mother — but they were fun-loving and musical, interested in dress, parties and entertainments.

The fact that the French capital had, since March 1832, been in the grip of an even more violent outbreak of cholera than London[4] did not worry Dr. Somerville. He was convinced, as many were, that the disease, for some unaccountable reason, never attacked the English in Paris.[5] Nor was the family perturbed by the possibility of civil violence. In June the French king had almost been toppled from his throne by rioting by the left in Paris and in the north the ultra-rightists, led by the Duchesse de Berri, mother of the Carlist heir, were arming for an attack on Louis-Philippe.

The Somervilles arrived in Paris in early September. Woronzow Greig, who visited them shortly afterward, wrote his grandmother:

. . . the attentions which my mother has received ever since her arrival are quite overpowering, every person of note who was then in town and whose acquaintance was worth possessing has waited upon her, her arrival was proclaimed in glowing terms in the newspapers and the French people generally seem to have been vying one with another who should shew her the most honour.[6]

The welcome extended by the Paris scientific community was cordial. Arago was Mary Somerville's first caller. Always a dedicated liberal, he had taken a leading part in the July Revolution that put Louis Philippe on the throne in 1830, now held several important posts, and was the most influential scientist in France. He and his colleagues, all of whom had long admired

Mrs. Somerville, now had a special regard for her since the English rendition of their revered master, Laplace. Laplace's old assistant, Bouvard, was also prompt in paying his respects and arranged a dinner party in her honour for 27 September.

But the flurry of the Paris reception proved too much. Mrs. Somerville had a relapse and took to her bed for a week. Her son and husband had no difficulty in persuading her that she should remain in Paris for the sake of her health and her daughters' education. Comfortable lodgings were found, a courier and cook hired, young Martha put in charge of managing the household, and what was to be a marvellous year abroad was underway. During the next months Mrs. Somerville and her daughters would move in many Paris circles — among the scientific and the fashionable, among English and American, among ancient French families and rising newcomers. They would meet distinguished men and women, French and foreign, and form lasting friendships. Moreover this Paris stay would have a profound effect on Mary Somerville's next work.

On 4 October[7] Bouvard's postponed dinner took place. Dr. Somerville wrote home that his wife 'performed her part at the Bouvard dinner, given entirely on her account, with a garnish of brains round his table — all the owners of which paid her the most marked attention'.[8] She herself sent more details:

. . . Mr. Bouvard's party . . . turned out remarkably well. The Marquise de la Place received me with open arms. She was the only lady present and did everything in her power to be kind and agreeable to me and the girls and will be a valuable friend as she goes to all the Queen's parties and is in the first and gaiest society in Paris. She came yesterday to invite us to dine with her on thursday [11 October]. All the scientific people were at Mr. Bouvards and I assure you I have reason to be vain of the flattering things that were said of me. I sat next Baron Poisson at dinner who urged me in the strongest manner to write a second volume and he afterwards did the same to Dr. Somerville and told him there were not above twenty people in France who could read my book. Mr. Arago was also there he said he had not written to thank me [for his presentation copy] because he had been reading the work and was busy preparing an account of it for the Journal of the Institute. Such opinions from the greatest mathematicians in Europe and from the astronomer royal of France are very gratifying. I was a good deal excited and fatigued with speaking french all day so that I felt languid the next morning. . . .[9]

Throughout her stay Mary Somerville would be part of this scientific circle, the most eminent in Europe. Madame Laplace, a woman in her sixties but 'still full of vivacity and amiability . . . cheerful, excellent and bright',[10] became a good and useful friend, intent on seeing to it that the Somervilles enjoyed their stay in Paris and profited from it. Her dinner was at Arcueil, 'the party not quite the same [as at Bouvard's] . . . & somewhat less scientific'.[11]

In 1824 when the Somervilles visited Brussels with Sir James Mackintosh, the British Ambassador, Lord Granville (1773–1846), invited them to dine;

the Somervilles were asked largely because of their distinguished traveling companion. Lord Granville was now Ambassador at Paris and Mary Somerville was a celebrity. The envoy and his wife, a daughter of the Duke of Devonshire, entertained the couple on the evening following Madame Laplace's dinner. Somerville proudly reported:

. . . Nothing could exceed Lady G's attentions. She offered me her arm to go to dinner which I had not presumed to think of. She desires your mother to introduce the girls to her, & I mentioned the probability of the three ladies remaining the winter without me under the care of our friends. She immediately said I hope you include us in the number & you may be assured we shall all be to happy & proud to take care of Mrs. Somerville. . . .[12]

Early in their stay the Somervilles met General Lafayette (1757–1834) and the two families became warmly attached. A vigorous and commanding figure at age 75, Lafayette, while out of favour with Louis Philippe, was still the most powerful voice in French politics, the most ardent spokesman for liberty everywhere, and to the Somervilles — with their liberal opinions — a legendary hero. They saw a great deal of him and his family, were twice at his famed country estate Lagrange,[13] and through him met the large American colony in Paris. The novelist James Fenimore Cooper was its leader and the Somervilles came to know him, his wife and their four daughters — roughly the age of their own two — very well.[14]

There were many English in Paris, and Mrs. Somerville and her daughters — after Dr. Somerville's recall to Chelsea late in November — were well attended. Their friend Patty Smith, wintering in Paris, introduced them to the Duchesse de Broglie,[15] a famed Paris hostess and daughter of Madame de Staël. A friendship quickly followed and the Somerville ladies were regularly at her weekly *salons*. Several other French hostesses proffered similar invitations. Lafayette and Arago — both members of the Chamber of Deputies — and its President, Baron Charles Dupin (1784–1873), whom the Somervilles had known well for over a decade,[16] on several occasions sent tickets admitting them to sessions of the Chamber of Deputies. There were frequent visits also to the opera and the theatre, to museums and galleries, and to various landmarks and places of interest. The three Somerville ladies were insatiable sightseers. They made jaunts outside Paris and visits to country estates. After the New Year, young Martha and Mary were presented at the French Court and embarked on the round of splendid balls and entertainments that filled the Paris season.

Mary Somerville was determined that her daughters should have every advantage of their Paris residence. The three ladies survived a formidably full ten months and survived with pleasure. Their long visit is described in several dozen gossipy letters home. These missives reveal not only the wide extent of their Paris acquaintance and the details of life in Paris in 1832–33 but disclose the glorious reception the admiring French gave Mary Somerville. 'How interesting it is to meet with so many nice people', she wrote her husband in the spring of 1833, 'we are certainly very fortunate in receiving so much kindness, and if ever we should return to Paris we shall be quite at home'.[17]

2. Mary Somerville and French Science, 1832–33

When Mary Somerville arrived in Paris in September 1832 she was already well along in the manuscript of her second book. She had no intention at the time of remaining for more than a short time in the French capital and no idea of the profound effect which a stay would have on her forthcoming volume. She had begun work even before the publication day of the *Mechanism of the Heavens* and had applied herself to it steadily. In her memoirs she explains that she had 'got into the habit of writing' and once the *Mechanism* was completed and revised, she 'did not know what to make' of her spare time. 'Fortunately', she declares, 'the preface of my book furnished me with the means of active occupation, for in it I saw such mutual dependence and connection in many branches of science, that I thought the subject might be carried to a greater extent'.[18] Others, including John Playfair, had sensed a connection, but Mrs. Somerville was the first to explore the matter using the most recent scientific findings.

The relations she perceived among the sciences were implied so fundamentally in her *Preliminary Dissertation* that a totally unmathematical reader, her friend Anne Napier of Reading, remarked in her letter of thanks for a copy of the treatise that she had been struck 'with the number of Sciences & natural facts, which are in close connection. This mutual dependence of all parts of truth upon each other is one of the most beautiful perceptions, which a glimpse at the happy world of Science presents'.[19] It was Mary Somerville's early recognition of these connections, together with 'the universal facility and occasional felicity of expression that distinguishes . . . [her] writing', that led James Clerk Maxwell (1831–1879; F.R.S., 1861), some forty years later, to cite her second book as one of those

. . . suggestive books, which put into definite, intelligible and communicable form, the guiding ideas that are already working in the minds of men of science, so as to lead them to discoveries, but which they cannot yet shape into a definite statement.[20]

This second book, *On the Connexion of the Physical Sciences,*[21] published in February 1834, incorporated, as Mrs. Somerville's correspondence reveals and its text confirms, the ideas and comments of many of her scientific intimates and of others whose work they considered important. In demonstrating the connections of the sciences, it set out the newest views obtaining at the time about the physical sciences. Unlike the *Mechanism of the Heavens,* which dealt with material long familiar to her — for she had begun her study of Laplace in the first decade of the century — the new book of necessity touched upon, as she confessed,

. . . many subjects with which I was only partially acquainted & others of which I had no previous knowledge but which required to be carefully investigated so that I had to consult a variety of authors British and foreign.[22]

Both in her preparatory reading and in her final writing, she was guided by scientific friends at home and abroad.

Among her papers, the earliest letter hinting at the new project is one written her in mid-March 1832 by W. I. Broderip.[23] In it Broderip supplies in detail some of the information she had requested about plants and animals in the Himalayas. This whole transaction — the mode of query, the careful formulation of an answer, and the use made of it — is typical of the way in which Mrs. Somerville gathered material for her books. Somerville had passed her enquiries on to Broderip as the two men 'sat in Council' at the Geological Society. Broderip's answer, dispatched the following day, was enlarged by a visit he had paid that morning to the British Museum to look over, on Mrs. Somerville's account, 'the Moorcroft herbarium preserved in the Banksian collection' with the Keeper of Botanical Collections at the Museum, Robert Brown (1773–1858; F.R.S., 1811). Broderip carefully referred her to Moorcroft's paper on the herbarium,[24] and to John Gould's (1804–1881; F.R.S., 1843) recently published book on Himalayan birds,[25] and — while sending her all the information he had on the subject — promised her 'further intelligence' as soon as possible.[26] All this information, and any other she gathered, she compressed into a 200-word passage on the effects of climate in the Himalayas on plants and animals there.[27]

Whewell and other Cambridge friends were also involved in the project from an early date. During the Somerville's Cambridge visit of April 1832, Mrs. Somerville discussed it with him and after her return to London wrote, in her letter of thanks for the hospitality,

. . . I had expected to avail myself of your friendly offer to look over some sheets within eight or ten days after our return . . . but . . . the fact is that matter has accumulated upon me so much, that the chief difficulty I experience is in limiting my notice of the subjects that fall within the scope of my plan. There is no help for it. I must go on as well as I can without hurrying which spoils everything. . . .[28]

To Mrs. Leonard Horner she confided a month later, 'I am as busy as this idle time [June] will admit of but I fear I shall not be ready to *print* this season'.[29] Nevertheless, in comparison with her *Mechanism*, this second book was very fast in writing. Mrs. Somerville's illness later in the summer, the journey to France, and her almost immediate relapse slowed progress. When Arago came to call soon after her arrival, she told him of her work and he immediately sent her 'some interesting memoirs, and lent . . . a mass of manuscripts with leave to make extracts, which were very useful . . .'.[30] His example was soon followed by others in the French scientific community.

Since her illness, she had fallen into the habit of writing in bed each morning until 1 p.m.,[31] then spending the rest of the day with her daughters and friends. In this way she from the first mantained a fairly rigid work schedule and also found time for a good deal of scientific interchange and conversation. The several dozen letters in the Somerville Collection written by her and her daughters during these months in Paris are devoted chiefly to gossipy accounts of their own doings and plans but there are many sketches of the persons they met and occasional glimpses of Mary Somerville's contacts with the French scientific community. Her memoirs give a much fuller listing of the scientists she saw in Paris in 1832–33. Very little correspondence between her and these colleagues is in the Collection, but the printed text of *On the Connexion of the Physical Sciences* shows how influential was their help.

The French philosophers she had met in 1817 were now the elder statesmen of science, men of position and authority. Many of the important figures in the Society of Arcueil — Arago, Biot, Gay-Lussac, Dulong (1785–1838; F.R.S., 1826), Poisson (1781–1841; F.R.S., 1818) and Thenard (1777–1857; F.R.S., 1824) — were still active, while several associates of this early group — Bouvard, Poinsot (1777–1859) and Prony (1775–1839), for example — were now recognized leaders of the French scientific establishment. Arago and Gay-Lussac — each still involved in a number of different projects — were now in the Chamber of Deputies, Bouvard was at the Observatory. Biot, in ill health, was often away from Paris at his country estate. All were delighted to welcome Mary Somerville again to France and to put at her disposal whatever facilities and information they had.

The Somerville letters suggest that Parisian society in 1832–33, while much freer, was much less mixed than was London society. In London the Somervilles were almost certain to encounter some of their scientific friends at any party they attended. Literary, political, scientific, and often fashionable, circles mingled at most London gatherings. There were on occasion, small groups of scientific friends that gathered to spend time together but even these little dinners and breakfasts and parties generally included friends of science and friends of friends of science. In Paris, where professionalization of science was much more advanced than in England, the Somerville letters reveal little mingling between the scientific sector and other circles.

In the surviving letters Bouvard is the French savant most frequently mentioned, and the relations described between him and the Somervilles during this long stay in Paris are characteristic of those obtaining between her and the greater part of the French scientific community in 1832–33. Not only did he entertain Dr. and Mrs. Somerville soon after their arrival in Paris — thus affording them opportunity to renew ties with a number of French savants and with Madame Laplace — but he continued to do so and was often among the guests at other scientific gatherings where they were present.

Two letters from him to Dr. Somerville, written in November 1832, are indicative of the kind of help Bouvard offered. The first is merely a brief note saying he had sent Somerville's letter to the chemist P. L. Dulong, who would call as soon as possible.[32] Dulong had just been elected permanent secretary for the physical sciences of the Académie des Sciences, where Arago served as secretary for the mathematical sciences. Bouvard's second letter — in answer to several questions about astronomy directed to him by Mrs. Somer-ville —lists quantities (some of them deduced from Bessel's calculations 'qui paraissent être jusqu'ici les plus exacts') for twenty separate astronomical phenomena and promises to send replies to her other questions as soon as he has the necessary information.[33]

In early April Mrs. Somerville and her daughters spent a day at the Observatory at Bouvard's invitation, an outing so pleasant that it prompted her to write her husband, 'How interesting it is to meet with so many nice people; we are certainly very fortunate in receiving so much kindness, as if ever we should return to Paris we shall be quite at home'.[34] The party Bouvard assembled for them that eleventh day of April included Madame Laplace, her genial and well-informed son, her grand-daughter with her young husband, a Mademoiselle de Rigny (who had been Bouvard's pupil and whom Mary Somerville came to like and admire), and the head of the Paris police and his wife.

The only practising scientist present, apart from the host, was Count Gustave Pontécoulant (1795–1874; F.R.S., 1833), whose work in astronomy Mrs. Somerville already knew. A graduate of the École Polytechnique, Pontécoulant, in the spring of 1833 when Mrs. Somerville met him, was in the process of turning from soldiering to a deeper involvement with mathe-matics and astronomy. He was one of the few younger French scientists Mary Somerville encountered during her stay in Paris. She wrote her husband:

. . . I do not think M. de Pontécoulant so taking in his manners but he is very agreeable and gentlemanlike and amused us with an account of the [1832] siege of Antwerp where he was aid-de-camp [sic] to General Gerard [who compelled the Dutch to leave that city].[35]

Mrs. Somerville included in her forthcoming book the date Pontécoulant, along with M. C. T. Damoiseau (1768–1846; F.R.S., 1832), calculated for the next appearance of Halley's comet,[36] due in 1835. It proved inexact and was removed from editions following the return of the comet.

Since the July Revolution, Arago had been largely occupied with politics and with national scientific and educational affairs, but he still found time for researches. Now chief director of the Observatory, permanent secretary of the French Academy, member of the Chamber of Deputies and of the Municipal Council of Paris, he was in a position to influence French science and was temperamentally and intellectually suited to the task. Under his impetus the works of Laplace and Fermat were published, Daguerre re-warded for his discovery of photography, the navigation of the Seine im-proved, the museum at Cluny opened, the building of railways and electric telegraphs encouraged. Mrs. Somerville in her letters mentions none of the projects then under way. Instead she writes of Arago the politician and says nothing of Madame Arago, who had been one of her principal hostesses in 1817.

In her book, however, there are frequent references to Arago and his work. The extracts which he permitted her to make from the memoirs he supplied were used to good purpose. There is much discussion of his earlier work with Fresnel (1788–1827; F.R.S., 1825) on polarization[37] and accounts of his recent conclusions about the correspondence of barometrical pressure and the phases of the moon,[38] the temperature of the earth,[39] and rotary magnetism.[40]

Biot had been Mrs. Somerville's chief sponsor on her earlier visit, but he was now in ill health and rarely went into society. He was impatient too with scientific colleagues who engrossed themselves with what he called the political nonsense of the day, and he and Arago were again on bad terms. Biot was engaged at the time in 'making experiments with light'[41] and in the midst of demonstrating dextro- and laevo-rotation of polarized light in solutions and solids. He sent her a copy, along with four or five others for English friends, of the paper he had read at the Academy in the early winter on 'la polarisation circulaire' and pointed out that he was continuing his investigations in the country.

. . . J'ai, [he wrote] dans les caractères de la polarisation circulaire un fil qui me guide à travers les merveilleuses opérations de la nature; et me m'étonne sans cesse des moyens simples qu'elle emploie pour varier tant de produits, dont nous étions loin de soupçonner la liaison.[42]

These findings made their way promptly into Mary Somerville's manuscript.[43]

When she told him of her new project, he was delighted and wrote:

. . . Un ouvrage tel que celui que vous m'annoncez, et fait par vous, ne peut manquer d'offrir beaucoup d'intérêt, et quant à moi, je pense qu'il est destiné à exprimer un fait d'une importance capitale dans l'état actuel des sciences. Maintenant, l'utilité du but sera-t-elle appreciée ici du premier abord; je n-oserais l'affirmer. Notre public se montre en général peu soucieux des considérations purement scientifiques. Il court volontiers après les spéculations paradoxales qui éveillent sa curiosité, ou après les applications qui solicitent son intérêt. Des oeuvres elles-mêmes, en tant que doctrines abstraites et sévères, il y est très peu sensible.[44]

He and his wife urged the Somervilles to visit them at their country home near Clermont-Oise, conveniently close to a return route to England, but it was not possible to accept their warm invitation.

There are passing references in the Somerville letters to the Gay-Lussacs. Mrs. Somerville had known them both in 1817 and Madame Gay-Lussac had sent her later a French nursery maid.[45] The Somerville ladies in December had to refuse a dinner invitation from the Gay-Lussacs in accord with the policy Mrs. Somerville had adopted after learning of her mother's death in

Edinburgh on 22 November. Lady Fairfax was in her ninety-first year, much beloved by her family, and in command until the end. She had imparted to her daughter her own intensely practical approach to life. Mary Somerville knew that 'under our present circumstances' her mother would approve of their not going into deep mourning and withdrawing from society. Instead she decided to accept no formal invitations in December — the Gay-Lassac's and one from the Poissons among them — but to go back into society in January wearing light mourning. As she wrote her son, not to do so 'would be defeating the object of our living here'.[46] Nevertheless she asked him not to mention the matter in London, fearing criticism.

On the last day of the year 1832 the three Somerville ladies paid a round of calls, as young Mary wrote her brother, including one to 'Mad^e Gay Lussac and her husband [who] is a Deputy and she told us a queer calculation'[47] about the amount each person in France would receive if all property were divided equally among the population. This story must have impressed the Somervilles for almost 50 years later Mary Somerville put it in her autobiography, again attributing it to Madame Gay-Lussac.[48] She also describes 'Madame . . . as being only twenty-one, exceedingly pretty, and well-educated; she read English and German, painted prettily, and was a musician'.[49] In all likelihood, Mrs. Somerville is remembering one of the Gay-Lussac daughters, for a few years later, writing to Martha Somerville of 'Madle Gay-Lussac's' marriage, she calls her a 'very pretty and accomplished girl'.[50]

Gay-Lussac himself was in 1832–33 completing the development of a series of volumetric methods of assaying that would, among other improvements, replace the old method of cupellation with a volumetric determination of silver. Mrs. Somerville does not mention this important work in her *Connexion of the Sciences,* since chemistry is not among her 'physical sciences'. There is no reference at all to Gay-Lussac in her first edition, but his name appears twice in the second (1835) and subsequent editions, attention being called to his discovery that gases unite in simple and definite proportions[51] and to his estimation of the length of a flash of lightning.[52]

Courtesy to Madame Gay-Lussac and to Madame Laplace[53] impelled Mrs. Somerville, according to her memoirs, to accept a distasteful invitation for 28 February 1833. On that date she and her daughters dined with these ladies at the table of François Magendie, who since the previous year had been professor of anatomy at the Collège de France. In the letters of the date, only passing reference is made to the engagement,[54] but in her autobiography, written after the antivivisectionist movement had gained strength and her outspoken allegiance, she deplores his coarse manners, 'horribly professional' conversation, and the fact that 'subjects . . . not fit for women to hear' were discussed during the evening'.[56] In the manuscript of her memoirs she had previously revealed that during the Somervilles' 1817 visit to Paris she had gone with her husband 'very unwillingly to dine with M. Magendie a man I absolutely abhorred for his cruel vivesection, they say he repented of them on his death bed'.[57]

In her letters Mrs. Somerville mentions dining with another scientist-deputy, Vicomte Hericart de Thury, who had been their visitor at Chelsea the previous July. Hericart de Thury, director of mines under Bonaparte and in charge of public buildings under the Bourbons, was a legitimist, his wife an ultra-legitimist who expressed horror when Mary Somerville revealed that she and her family had stayed with the Lafayettes. Madame de Thury implored her not to mention the fact aloud lest some of the other dinner guests — '2 Peers and Peeresses & un Marechal de France and some others'[58] — leave the room. The lady received every Monday during the season and the Somervilles occasionally accepted the standing invitation she proffered them.[59] Political feelings entered the drawing-rooms constantly during the season, the Carlists were adamant in their refusal to be civil to their opposites,[60, 61] while even the moderates easily became overwrought.

Inexplicably there are few references to Poisson in the correspondence of the period. Mrs. Somerville admired him greatly and made much use of his work in her studies and her writings. In the recent past she had discussed his findings, corresponded about them, and exchanged the Frenchman's papers with J. W. Lubbock in London,[62, 63] and continued to do so while in Paris. Lubbock in December 1832, responding to a recent communication from her, ended his letter with the words, 'Mr. Poisson is indeed an example which many here might follow with advantage, you have done so much to put them to shame. I hope you will do much more'.[64] Furthermore, Madame Poisson — who spoke English, having lived in that country with her family during the French Revolution — had kind recollections of seeing Mrs. Somerville at Madame Biot's in 1817[65] and was warmly hospitable to her on her return to Paris in 1832. A note from Poisson, dated 28 November 1832, for example, reminds Mrs. Somerville that she and her daughters are to dine with them on the following Monday,[66] an engagement that had to be cancelled after news of Lady Fairfax's death reached Paris. Poisson and Mrs. Somerville found a wider community of interest than mathematical astronomy; one draft of her autobiography recounts Poisson's illustrations of how the 'inclination of girls was little considered' in making marriage arrangements in France,[67] an observation on French customs and French character that Mrs. Somerville took as authoritative.

She was flattered that 'Baron Poisson' — a title he received from the Bourbons in 1825 but never used — urged her, first by letter[68] and then in person,[69] to continue her work on Laplace and began to do so while in Paris. In her memoirs she reports that

. . . in consequence [of following Poisson's advice], I was led into a correspondence with Mr. [James] Ivory who had also written on the subject [of the form and rotation of the earth], and also with Mr. Francis Baily, on the density and compression. My work was extensive, for it comprised the analytical attraction of spheroids, the form and rotation of the earth, the tides of the ocean and the atmospheres, and small undulations.[70]

Mrs. Somerville's comments on French science in letters to her husband

and son at this time are few and superficial. Neither there nor in her
autobiography does she refer to one of the principal differences then obtaining between French and English science: the attitude toward professionalization prevailing in each country. French scientists had become increasingly
more professional over the past half century and by the 1830s this change was
manifested in increasing numbers of doctoral candidates in scientific subjects,
in an established system of science instruction that offered many new classes
of junior posts, and in heightened opportunities to young scientists for
continued state employment.[71] Many of Mrs. Somerville's closest French
associates had long profited both in their livelihoods and scientific work from
national support. In Britain, on the other hand, the push toward professionalization had barely begun. Mary Somerville in 1830 was at the center of
efforts by reformers to assert a doctrine of professionalism over patronage,
and the issue which had so agitated the Royal Society in that year was far
from forgotten in 1832–33. Yet whatever discussion — if any — which she
had with French philosophers about the state of science in England and
France, whatever views she may have developed from observation of the
strengths and weaknesses of professionalized French science and of the strong
national scientific establishment, whatever comparisons she might have
drawn between current French and British practices — all are unexpressed.

Nor, seemingly, did she notice the rising generation of French savants. In
London she was always in touch with veteran natural philosophers and with
bright newcomers to science, but in Paris her associations in 1832–33 appear
to have been almost wholly with the survivors of the group that in 1817 had
surrounded Laplace and Berthollet. Only three names not closely identified
with the Society of Arcueil occur in her accounts of her second visit — and
these in the autobiography, not in the letters. All three are names of men of
her own generation: William Frederick Milne Edwards (c. 1776–1842;
F.R.S., 1829), A. M. Ampère (1775–1836), and A. C. Becquerel (1788–
1878; F.R.S., 1837).

In her printed memoirs Mary Somerville declares, 'No one was more
attentive to me than Dr. Milne-Edwards, the celebrated natural historian'.[72]
She says nothing of the fact that in November 1828 her husband, together
with Dr. John Bostock (1773–1846; F.R.S., 1818) and J. F. W. Herschel,
had signed Milne-Edwards' certificate for fellowship in the Royal Society.[73] In
the first manuscript draft only of her autobiography, she describes 'Dr. Edwards' — as he preferred to be known — as 'conceited but he was the first
Englishman elected a member of the Institute'.[74] He and his younger brother,
Dr. Henri Milne-Edwards (1800–1855; F.R.S., 1848), also active in French
scientific circles, were useful to many visiting English scientists, for they were
bilingual. The sons of a wealthy Jamaican planter, they had grown up in
Brussels, had been educated there and in Paris (where both had taken
medical degrees) and pursued distinguished careers in physiology, anatomy,
natural history and related studies. The elder had worked with Magendie in
1816–17 and in 1824 published an important book on the influence of
environment on animal economy.[75] Mrs. Somerville relates how she procured

for him certain information from Sir John Sebright on acquired instincts in wild ducks.[76] In 1832 he was elected to the Académie des Sciences and in the same year admitted to the French Institute as head of the recently-created class of moral and political sciences. In the next years his interests would turn to a new study, ethnology, whose creator he largely became. It was to him, Mrs. Somerville declares, that she 'was indebted . . . for the acquaintance of MM. Ampere and Becquerel'.[77]

Ampère had been on the fringe of the Arcueil group,[78] Becquerel hardly that close. Yet both by their work in electrodynamics had complemented and extended studies by Biot, Fresnel and others. Ampère formulated a mathematical theory of electrodynamics to which Mary Somerville devoted two full pages in her printed book, remarkable attention at the time. He had, she declared, brought 'the phenomena of electro-magnetism . . . under the laws of dynamics . . .'.[79] Becquerel since 1829 had been experimenting with electrical phenomena, and she reported his success in forming mineral crystals through voltaic action in her first edition.[80] In her second edition she added accounts of his work on electrical phenomena,[81] his theory of atmospheric electricity,[82] and his thermoelectric battery.[83] All these studies were at the time very new; Becquerel's first publications of these researches came in 1834.

All other French work cited in her *Connexion of the Sciences,* however, is overwhelmingly from the researches of the Arcueil group. Mrs. Somerville's assessment of French scientific merit, her selection of French material to be included in the volume, and the interpretation to be placed upon it are all clearly guided by the opinions of Laplace's old circle. In these philosophers she had trustworthy guides, but once they were gone she had little direct access to French science.

Paris, though filled with cultivated and accomplished women, had no set of 'scientific ladies' comparable to the group found in London. Madame Biot had assisted her husband early in the century and no doubt other wives blended their efforts into those of their spouses, but science was not the fashion in Paris as it was in London. Madame Laplace had been the premier scientific hostess since the days of Arcueil, Madame Arago had assumed a prominent place after her husband became secretary of the Académie des Sciences.[84] Mrs. Somerville, in her letters of 1832–33 and later in her memoirs, mentions only Mademoiselle de Rigny, a woman of her own age and niece of the famed finance minister, Baron Louis (1755–1837), as being scientifically or mathematically inclined. Orphaned by the Revolution and left in charge of four younger brothers, she had superintended their education, even taught them herself. One of them, Comte Henri G. de Rigny (1782–1835) was in 1832–33 Louis Philippe's minister of naval affairs and would become foreign minister in 1834. Mrs. Somerville considered Mademoiselle de Rigny and Madame George Washington Lafayette 'the cleverest and most agreeable french women we have ever met with'.[85] The Somervilles spent a glorious Sunday in May at Petit Brie, Baron Louis' beautiful estate in the Bois de Vincennes, where Bouvard and his nephew Eugene were also guests.

Mrs. Somerville described to her husband how much at home she felt on entering its drawing-room with 'the last numbers of the Edinburgh and Quarterly Reviews on the table'.[86]

Her experiences with two other female connections of French scientists were not so happy. When the Somervilles visited Paris in 1817, the Cuviers received them graciously, entertained them, and since that time had maintained friendly relations. When Mademoiselle Clementine Cuvier — the last survivor of the couple's four children — died in 1827 at age 22, Mary Somerville sent her condolences,[87] remembering no doubt her own bitter loss of a daughter. In 1830 when Cuvier visited London, he was accompanied by his step-daughter, Mademoiselle Duvaucelle,[88] whose charms — along with those of his beloved Clementine — had attracted visiting intellectuals from the provinces, scientific foreigners and such literary men as Stendhal (1783–1824) and Prosper Mérimée (1803–1870) to the Cuvier *salon*.[89] Cuvier himself died on 13 May 1832, some five months before the Somervilles returned to Paris, but his widow and her daughter continued to occupy their old quarters at the Jardin des Plantes. Mademoiselle Duvaucelle called promptly on the visitors.[90] On 15 November the Somervilles spent an interesting day at the Jardin des Plantes, watching the animals[91] and dining pleasantly 'with Mme C and a little party of gentlemen'.[92] In the spring, however, relations between the two families became strained when a box, dispatched to London by Mademoiselle Duvaucelle through the Somervilles, failed to reach its destination. In the first draft of her autobiography — it is omitted in all others — Mrs. Somerville wrote that 'sometime afterwards when I called I was very ill received. I never met with more pretensions & insolence from any woman than I did from the niece Mlle de Vaucelle . . . Of course I cut the acquaintance'.[93] The breach must have been deep for there is no further mention of the Cuvier family. The incident, even the mistaken relationship, represents a rare termination in good feelings between Mary Somerville and others.

Madame Rumford, twice a scientific widow, maintained in 1832–33 one of the great *salons* of the capital. She was old, rich and disagreeable. Her first husband was the chemist Antoine Lavoisier (1743–1794), whom she sometimes assisted by drawing plates, making notes, and translating from the English.[94] After his death she edited and printed his fragmentary *Memoirs de Chimie* (1805). In 1804 she married Benjamin Thompson, Count Rumford (1753–1814; F.R.S., 1792); their union was stormy. During the Paris season Madame Rumford received each Friday evening at her sumptuous residence in the Rue d'Anjou St. Honore. At first she greeted the Somervilles 'very kindly'[95, 96] and they were pleased to be noticed. After a few weeks her notoriously bad temper began to assert itself, and by the end of February young Martha was describing her as a 'nasty old hag'. She wrote her brother in London, 'I sometimes think she is jealous of Mama's meeting attention from people but whatever it is we are not friends'.[97] Nevertheless Mrs. Somerville continued to go to the Friday receptions on occasion to hear news of visitors to Paris.[98]

In March 1833 the fashionable Paris sculptor, Pierre David d'Angers (1789–1856), asked, through Arago, if he might model Mrs. Somerville's profile for a medallion. Her scientific friends wished to have some likeness, and David considered her a timely subject. Two sittings were sufficient[99] and by late April the medallion was cut and delivered.[100] On the whole it was thought 'very like',[101] but in bronze some considered it '. . . very hard . . . the Bumps of Mathematics . . . too strongly marked . . . [the whole] too cut out, too unfeminine . . . but [the cost] not extravagant'.[102] David presented Mrs. Somerville with 'a medallion in bronze, nicely framed, and two plaster casts for . . . [her] daughters'.[103] The medallion became popular among her admirers and is today one of her best-known likenesses.

Busts and medallions were frequently exchanged in these decades as evidence of esteem and good will. A request for a likeness signalled high regard for its subject, its presentation high regard for the recipient. Late in June 1832, Madame Laplace, wishing to show honour to her late husband's American translator, Nathaniel Bowditch, dispatched a bust of Laplace to America for him.[104] On Mary Somerville's arrival, the mathematician's widow decided to make the same present to her.[105] When word reached England in mid-October — through Woronzow Greig — of the intended gift, the Royal Society, by way of its treasurer J. W. Lubbock and through Greig, sent an informal request to Mrs. Somerville: would she kindly suggest to Madame Laplace that the Society would be honoured to add to its small but exclusive collection of likenesses a bust of the great Laplace? Mary Somerville gladly made the request,[106] Madame Laplace gladly assented,[107] and in May the bust — together with the one for Mrs. Somerville — crossed the Channel.[108]

Lubbock had been for several years one of the Somervilles' younger circle. Two years Greig's senior, they had been at Trinity College, Cambridge together from 1823 until 1825, when Lubbock took his degree and was first senior optime. After travels on the Continent, the young man entered his father's banking business in London. The interest and proficiency in mathematics he had demonstrated at Cambridge continued and he soon found a place in the London scientific community, becoming a member of the Astronomical Society in 1828 and of the Royal Society the following year. In the 1830 election he was one of the few members of the reform faction to gain office in the Society, being chosen Treasurer, a post he held for the next five years and then later from 1838 till 1845.

He and Mrs. Somerville had much in common. Introduced to Laplace's mathematical techniques during a visit to Paris in 1822, Lubbock began to apply Laplace's probability theory to annuities,[109] to lunar theory,[110] and to the study of the tides[111] in the years that Mary Somerville was engrossed in her Laplace rendition. In conversation and through correspondence they discussed various mathematical methods and findings, astronomical observations and theories, and their own researches.[112, 113] Each requested the other to read, correct and criticize work in preparation.[114, 115] By the time the Connexion was published, the two were firm friends and colleagues.

Mary Somerville was pleased to be Lubbock's intermediary in the matter of the Laplace bust for the Royal Society in late 1832. A few weeks later, however, she learned with dismay that he had 'been obliged to give up his canvass for Cambridge . . . [for want of] enough liberal men in the university to carry it' in a by-election. She attributed his failure to 'the tottering state of the Church' but predicted that 'they [the university conservatives] will have some difficulty in supporting the falling edifice'.[116] Lubbock, replying to her 'kind remarks with respect to the Cambridge Election', confessed that, 'Until I went amongst them I had no idea of their contempt for science, I speak of nearly all without the walls of Trinity College',[117] known as a Whig stronghold[118] despite the fact that its Master, Christopher Wordsworth (1774–1846), was a High Tory.[119] A few months later, Mary Somerville rejoiced at news of Lubbock's approaching marriage.[120]

She was pleased earlier to hear that her own son had been elected on 7 February 1833 a Fellow of the Royal Society.[121] Young Greig, who had been admitted to the Inner Temple in 1824 and called to the Bar in 1830, was a barrister in London and already making a name for himself on the Northern Circuit. He had spent a good part of his life among scientific folk and had an interest in science but no inclination to make it his life's work. A singularly sweet-tempered, energetic and intelligent man, he was warmly regarded by his mother's friends as a tower of strength to the Somerville family. Several of his mother's close scientific intimates — Brewster, Lyell, Murchison and Chantrey — signed his certificate, as did his stepfather, William Somerville. A medical man — Sir James McGrigor — who had signed William Somerville's certificate fifteen years earlier also put his name to young Greig's, as did the famed army surgeon S. D. Broughton (1787–1837; F.R.S., 1830) and the physician and chemist John Bostock. Other old family friends — the respected Tory politician Sir Robert Harry Inglis (1786–1855; F.R.S., 1813), the classical topographer and numismaticist W. M. Leake (1777–1860; F.R.S., 1815), and the orientalist and numismaticist William Marsden (1754–1836; F.R.S., 1783) — endorsed the application, along with a distinguished barrister, David Pollock (1780–1847; F.R.S., 1829); and a recently resigned Fellow of Trinity and Greig's good friend, H. P. Hamilton (1784–1880; F.R.S., 1828). The name of Ralph Watson (F.R.S., 1830) is also affixed. Of the seventeen signers only one, W. H. Wollaston's nephew Alexander Luard Wollaston (1804–c. 1895; F.R.S., 1829), was of Greig's own generation. Although never a practising scientist, the new Fellow was a faithful member of the Royal Society until his death. His foster brother, Dr. James Craig Somerville, on the other hand, never became a Fellow.

By the time Mrs. Somerville returned to London, Greig had enjoyed nearly five months of his new scientific association. She had enjoyed, on the other hand, ten stimulating months in Paris. There she renewed old ties with many distinguished French philosophers, made several new scientific acquaintances, and — most important of all — became so familiar with current ideas and findings in French science that she was able to bring them to England in her forthcoming *Connexion of the Physical Sciences*. What news she brought the French about English science in this interval lies unrevealed.

3. Foreign Visitors, English Correspondence

During her months in Paris Mary Somerville was never out of touch with her English scientific colleagues or with her publisher, John Murray. There were also scientific visitors to Paris, some of them old friends, some of them strangers but all eager to see Mrs. Somerville.

One of her first visitors was the American medical student, Henry I. Bowditch (1808–1892). His father Nathaniel wrote him on 16 September 1832 of the favourable account of her *Mechanism of the Heavens,* coupled with the notice of Bowditch's Laplace translation, in the July number of the *Quarterly Review.* When the young man heard from Madame Laplace that Mrs. Somerville was in Paris, he — anxious to meet her but without a letter of introduction — went along confidently, as he wrote his father, 'as the son of the American translator of Laplace' to call. She received him 'very graciously and told . . . [him] that Lafayette had promised to introduce them, as Mrs. Somerville wished to sent a letter to the senior Bowditch. The description he dispatched to Boston is a vivid one:

. . . She is a lady of about common size, and has a thin, pale, and at the same time intelligent face. There was an air of mildness, amiability, and of modesty with regard to her own powers, which was very pleasant. Her character as described to me by those who know her corresponds with her general appearance, being amiable, domestic and although possessing an extraordinary mind, seems conscious of how little man can know. She has a pleasant, mild voice, and speaks with a Scotch accent.

She intends spending the winter here with her two daughters, who are about sixteen and seventeen, for their education. During my conversation with Madame, a tall stout gentleman entered, and was introduced as Dr. Somerville of London. He also greeted me cordially, promised what he would do for me when I arrived in London, and of the letters he would give to the first physicians in Edinboro'. After spending a quarter of an hour with them, I took my leave. Madame, shaking my hand, said she hoped to see me during the winter . . . In relation to the letter which she intends sending you, father, it is to thank you for the present of your translation. She spoke of it highly.[122]

When Bowditch did travel to London in the summer of 1833, Mary Somerville through letters and William Somerville through introductions made it possible for him to meet a number of scientific and medical persons in England and Scotland and to see various hospitals and medical establishments. He was always grateful for these kindnesses. At the time he wrote his parents of a dinner at Francis Baily's on 8 August at which Dr. Somerville and Charles Babbage were also guests and mentioned meeting Airy, Herschel, and Sir Astley Cooper[123]

Months earlier, in February 1833, Dr. Somerville had entertained another medical man, the colourful Clot Bey, Inspector General of the Egyptian Health Service. Antoine Barthelmy Clot (1794–1867) was a French doctor who in 1825 entered the service of the fierce and wily Albanian viceroy of Egypt, Mehemet Ali (1769–1849). Clot skilfully reorganized Egyptian medical services, establishing hospitals and medical schools. In 1832 Mehemet dispatched him on a special mission to Louis Philippe and the doctor on his return to France found himself lionized. His history, his high position under a foreign despot, and the dazzling diamond stars and crescents that decorated his uniform all attracted great attention at Parisian balls and receptions; the Somerville girls wrote excitedly of seeing him.[124, 125]

In February he made a brief trip to London, during which Somerville showed him over Chelsea Hospital,[126] and an even briefer one to Cambridge, where he was the guest of Whewell and others at Trinity College.[127] Within a few days of his return to Paris, he called on Mary Somerville and came again just before his departure for Egypt a bit later. Through him she sent a copy of her *Mechanism of the Heavens* to Mehemet, commenting wryly to her son:

. . . I had a visit of the Chevalier Clot-Bey on his return from England [around 26 February]. I never saw a man more delighted with a country and the reception he has met with than he seems to be with ours, he certainly is a very remarkable person, and agreeable withal. He is mightily pleased to take a copy of my book to the Viceroy of Egypt, it is curious enough that the Emperor of Russia [Nicholas I (1796–1855)] should be the only other potentate to whom it will be presented, barbarians both.[128]

Sometime after his return to Egypt, Clot-Bey wrote Dr. Somerville, 'Le livre de Madame de Sommerville [*sic*] fixe particulièrement l'attention de son altesse. Elle s'étonna qu'une dame eût entrepris d'écrire sur une matière aussi abstraite'.[129] He thanked Somerville effusively also for his kindness to him in London and expressed the hope of meeting the couple again.

Mary Somerville did meet again, in May 1833, another of their earlier London visitors, James David Forbes, who had just returned to Paris to resume a holiday interrupted by circumstances which he now sought to explain to the celebrated scientific lady. In February 1833, the Town Council of Edinburgh by a vote of 27 to 9 elected Forbes — not yet twenty-four years old — to the chair of natural philosophy at the University of Edinburgh. He was chosen over four other and much senior candidates — his old mentor Sir David Brewster, Professor William Ritchie of London, Professor Stevelly (1794–1868) of Belfast, and Thomas Galloway, mathematics master at Sandhurst and reviewer of Mrs. Somerville's *Mechanism*. Brewster, whose relations with several important English scientists had deteriorated over the past three years,[130] was especially chagrined, for numbers of his English colleagues had endorsed Forbes. The young philosopher was in Paris when John Leslie died on 3 November 1832 but hurried back to Edinburgh, where his family had already put his name in nomination as Leslie's successor. Family influence, political partisanship, and personalities played as much a part in the ensuing election as did scientific merit. The choice of such a young man caused a good deal of talk in British scientific circles — where such desirable posts were few — talk of which Forbes was aware and to which he responded indirectly in print.[131]

Some three months after the election, he returned to Paris, where, soon after his arrival, he became ill of a fever and appealed to Mrs. Somerville to suggest a doctor. She sent her neighbour, 'Dr. Morgan', in whose care Somerville had placed her own health[132] and who had assisted them when their young French maid had a nervous breakdown.[133] Forbes was soon better, and on 19 May sent her a long letter of thanks which dealt primarily with the recent Edinburgh election:

Several months ago I forwarded to London to you, a copy of the testimonials which I circulated on a late occasion. As I think it being probable that it never reached you, I venture to send you another, & I do this even at the considerable risk of appearing chargeable with vanity, upon the same grounds which have already induced me to give these considerable circulation, namely, in my own defence, & still more those of my Electors. The *prima facie* evidence in favor of Sir D. Brewster was so obvious, that it is surprising that persons at a distance, (who I find in some instances have been ready enough to pre-judge the question) should not have enquired as to the state of feeling near the field of action among those best qualified to judge. Had they done so, they would have found that there were considerations independent of Sir D. Brewster's scientific claims (which no one so readily acknowledges as myself) which made *some of his most intimate* & most Scientific friends *my best supporters*. I need only mention the President and Secretary of the Royal Society of Edinb — Sir Thomas Brisbane & Mr. Robison [later Sir John Robison (1778–1843)] whom everyone who knows Edinburgh must be aware are among Sir D.B.'s *most particular* friends & were the best qualified to judge the duties of the Chair, — My excellent friend Mr. Robison particularly so, since his distinguished father so long filled it — and it was upon the opinion of these men, & such as these that the Electors made [up = crossed out] their [opinion = crossed out] choice — some I know were guided very naturally by the opinions of the Professors (and there I speak within bounds when I say that I had *three* to *one* in my favour) — and yet these men, acting on no conceivable motive but that of rectitude, against a clamour, have been calumniated in a way which I shall not characterize. I do not blame Sir D. Brewster. My opinion is unchanged, for I have never shut my eyes to his weak points, which we all must have, — but I *do* blame his most injudicious friends.

Pardon, my dear Madam, this long explanation of my reasons for troubling you with the pamphlet. I cannot help feeling somewhat keenly on the subject on which I and my friends have been misrepresented — so much so that I found it current in several quarters that Sir D. Brewster and I had quarreled which is absolutely contrary to fact. I have the satisfaction (the best one can have in this world) of [having = crossed out] looking back on the affair with the consciousness of having acted straightforwardly & for the best, & that were I again placed in such difficult circumstances, my conduct would be regulated on precisely the same principles.[134]

Clearly Forbes wished to persuade the influential Mrs. Somerville, as he wished to persuade others, that his election was justified on the grounds of recognized scientific merit. The fifty-one printed testimonials in his pamphlet included those of several close Somerville friends — Herschel, Buckland, Whewell, Airy, Lyell and Murchison — and of others who had also been among the reform group in the Royal Society — Peter Barlow (1776–1862; F.R.S., 1823), Thomas Brisbane (1773–1860; F.R.S., 1810), William Henry

(1774–1836; F.R.S., 1809), William Prout (1785–1850; F.R.S., 1819) and Edward Turner (1798–1837; F.R.S., 1830) — as well as traditionalists such as J. G. Children (1777–1852; F.R.S., 1807). Some of the testimonials came from men instrumental in organizing the new British Association — John Phillips (1800–1874; F.R.S., 1834), Charles Daubeny (1795–1867; F.R.S., 1822) and William Scoresby (1789–1857; F.R.S., 1824) among them —while others were from men who had little concern with scientific groups. Mary Somerville, with her Scottish ties with Brewster and his family, with her high regard for his scientific ability and her firsthand knowledge of his personality, and occupying as she did a unique place in London scientific life, could be for Forbes an important ally or even an important neutral in the ticklish situation he now faced. Her acceptance of the electors' choice would be useful to him in Edinburgh and in London, and he plainly wished her approval. She, on the other hand, seems to have remained aloof from Edinburgh squabbles and kept on good terms with Brewster and with Forbes.

Early in their Paris stay the Somervilles met another colourful foreign visitor, Prince Koslofski. Martha Somerville described him at the time as a 'dumpy squat cossack . . . a very clever man with a prodigious memory'.[135] He lived in their Paris street,[136] and soon he and his young daughter Sofka, aged 16, became favourites with the family. In her autobiography Mrs. Somerville recalled they had

. . . met with Prince Koslofsky every where, he was the fattest man I ever saw, a perfect Falstaff and no wonder for he would eat a leg of mutton or a whole turkey besides fish & other things. However his intellect was not altogether smothered for he would sit an hour with me talking about mathematics, astronomy, philosophy & what not. He had offended the Emperor [of Russia], & was banished from Russia.[137]

Koslofski had written an elementary book on conic sections for the 'little Imperial Majesty of Russia', later Alexander II.[138] Mary Somerville says nothing of his relationship with the beautiful English-born Frances Sarah Lovell, wife of the Italian Count Guidoboni Visconti, who also lived in their Paris street. Countess Visconti's young son was, in fact, Koslofski's. Both she — who later became Balzac's mistress — and young Sofka Koslofski visited the Somervilles at Chelsea in 1835.[139] Mary Somerville maintained a correspondence with Prince Koslofski for some years.[140]

During the season there were a number of British scientific visitors in addition to Forbes. The Lyells arrived in mid-June on their way to visit the Leonard Horners in Bonn.[141] Jane Marcet stopped *en route* to Switzerland.[142] Their friend J. B. Pentland (1797–1873), the explorer and geographer, was often in Paris.[143] At no time were the Somerville ladies cut off from London scientific news and scientific gossip. Dr. Somerville at Chelsea was much in scientific society and always ready to carry out scientific commissions.

One of the letters directed to Mrs. Somerville early in 1833 is of particular interest not only in its scientific content but because it offers rare insight into one of the more prickly mathematicians of the day, James Ivory (1765–1842;

F.R.S., 1815). Like Mary Somerville, he was an exponent of French analysis but, beginning at the turn of the century, he had found little appreciation in England of his work. Even in 1833, after three decades of effort and the award to him of two gold medals (the Copley in 1814 and the Royal in 1826), a pension, and a knighthood, he still felt — in sharp contrast to the Somervilles — isolated from the inner circles of the London scientific elite. His letter not only lays out at some length his views about the figure of the earth but goes on to point out deficiencies in the work of others, opposition to his own labours, and popular denigration of 'impractical' studies in contrast to 'practical' endeavours. Ivory, it was whispered, neglected his teaching duties to pursue the higher branches of science, and the concluding passage of his letter reveals his resentment of these charges:

> After all what are the reasons for the continual gasconading about practical knowledge and practical men. Nobody that I know of, has under-rated the one, or made any attempt to lower the estimation in which the other are deservedly held. On the contrary practical men seem to have been liberally rewarded. There is so much vehemence in the declaration constantly issuing from the Press on this point, that one may reasonably suspect interested views, or vanity and the insatiable desire of personal distinction. But, as the saying is, the hotter the war, the warmer the peace; the greater the fury with which the indefatigable gentlemen blow the Fire of their zeal, the sooner will the good sense of the Public bring it down to a just temperature. One thing is certain and may be proved by indisputable facts, that the kind of knowledge not susceptible of being generally diffused, ought either to be entirely given up, or cultivated with more care.[144]

This crabbed mathematician apparently found in Mary Somerville a colleague to whom he could open his heart.

The English astronomer Francis Baily sent her, by way of William Somerville, answers to some of her 'enquiries, relative to the compression of the Earth, as deduced from geodesical measurements, from experiments on the pendulum'. Baily assured her that he would 'at all times, be most happy to communicate any information' in his power. He gave Dr. Somerville the good news that 'the Council of the R. Astronomical Society have ordered a copy of the *Greenwich Observations* to be, from time to time (as they appear) given to Mrs. Somerville' and inquired whether the eight parts already printed might be sent to the Athenaeum Club, 'as you may probably find means more rapidly of getting them from that place'.[145] The Athenaeum Club served in these years as the informal centre of British scientific life, a convenient location where messages and packages could be dropped and called for, where easy scientific congress was always possible. The *Greenwich Observations* were not readily made available in this fashion, and the fact that the Royal Astronomical Society had chosen to send this important series to Mrs. Somerville indicated in what high opinion they held her work. The *Observations* were of great benefit to her and her books.

Two other pieces of scientific correspondence are worthy of notice. The first is from a mathematician struggling to establish himself, the second from a philosopher already eminent. Early in 1833 Thomas Stephens Davies (1795–1851; F.R.S., 1833) sent Mrs. Somerville in Paris some of his latest mathematical researches and promised offprints of two forthcoming papers.[146] A Scotsman supporting himself in Bath by scientific and mathematical writing, Davies had come to her attention through Robert Ferguson of Raith and won her approval when he sent her some interesting and inventive work he was doing in spherical geometry.[147] After her kind reception of his work, Davies began to look upon Mrs. Somerville and her husband as scientific patrons. Already (1831) a Fellow of the Royal Society of Edinburgh, he was eager to be made one in the Royal Society of London as well. In 1832 William Somerville, along with Robert Ferguson, S. P. Rigaud (1774–1839; F.R.S., 1805), Baden Powell (1796–1860; F.R.S., 1824), Peter Barlow and a few others, signed his certificate but some inaccuracy in its execution delayed matters until January 1833 and Davies was not elected until mid-April of that year.[148] In 1834 he became a mathematics master at Woolwich. Never in the forefront of British mathematics or part of the central Scientific Establishment, Davies was nevertheless a sound and imaginative scholar whose worth Mrs. Somerville quickly recognized. An almost instinctive appreciation of quality and a willingness to support it wherever and in whatever circumstances was characteristic of Mrs. Somerville throughout her life. In the same fashion William Somerville was always willing to be helpful in matters such as Royal Society certificates.

William Whewell, busy arranging the third meeting of the British Association, sent Mrs. Somerville in April a pressing invitation to attend. His letter not only reflects some of the anxieties that beset the organizers of these early meetings but makes clear the fact that women — especially scientific women — were cordially welcome at the annual getherings:

I dare say you recollect that we hold a meeting at Cambridge this summer when we hope to see all our friends who take any interest in the progress of knowledge; and you would easily guess without our saying it, that we should be especially delighted to see you there on such an occasion. Dr. Somerville tells me that he does not know whether you will return to England before the time of our meeting (the 24th of June next). If you are at all hesitating as to the time of your return pray decide in such a manner as to give us the gratification of your company at that time. I hope we shall have most of the eminent scientific persons in England and several foreigners, and we shall do all we can to be both wise and amiable, characters in which we are especially desirous that you should see us. I expect Mrs. Buckland and Mrs. Murchison and several other ladies, and am to take steps for their accommodation as I return to Cambridge; I wish much I might also have the pleasure of making provisions for your reception.

I will send along with this some copies of our circular letter which if you can without inconvenience put them in the hands of some of the French men of Science may serve to shew them that we shall be extremely glad to see any of them who will visit us.

I take the liberty at the same time of sending a few copies of my Report on Mineralogy drawn up for the last meeting of the Association. I should be much obliged if you could without too much trouble have them conveyed to [F. S.] Beudant [1787–1862, a mineralogist], [Frederic] Cuvier [1773–1838; F.R.S., 1835, a zoologist], and [Alexandre] Brongniart, and to the Geological Society of Paris.[149]

Whewell's letter suggests the wide range and high calibre of French scientific associations her British colleagues expected Mrs. Somerville to have in Paris. Whewell also mentions in his long letter another of his own recent works and her coming book:

I have just published my Bridgewater Treatise[150] which I could wish you much to see — partly because I have had to go over some of the general views of the bearings of science which though different from those which you will probably have to present, have some connexion with them. I sent a copy to Dr. Somerville for you, but I think I shall request him to send it to Paris on the chance of your finding a spare moment for it now and then. He tells me that he will send me your "exposition" to peruse as it passes through the press which I will do with great pleasure and interest.

I do not know whether the connexion of science with Natural Theology is likely at all to interest any of your French acquaintance. My book appears to have a share of popularity here; if any one at Paris wished to translate it I should be glad to supply him a copy.[151]

Mary Somerville herself was anxious to read Whewell's book, fearing it would cover much of the same material as her own. Before the month was out, however, she could write her son:

. . . I am glad my book is in the press at last and I am still happier to find that Whewell's differs too much from mine to knock out my brains. He wishes to have it translated into French and has written to me on the subject. I shall enquire about it and let him know. I had some idea of the same kind with regard to mine but I shall take no step till I hear from your or Dr. S.[152]

She had some reservations, nevertheless, about certain aspects of Whewell's treatise which she voiced in private:

. . . I am rather angry at him [Whewell] for joining in the hue and cry against mathematicians for irreligion; a vulgar and monkish prejudice. All the philosophers were unbelievers before, and during the french revolution, the time when mathematics were cultivated with the greatest success, the fault is owing to the period, and not to the pursuit. I have no doubt however that it will make the book very popular among the Saints. . . .[153]

While never making a public issue of her disapproval, Mary Somerville always regarded with aloof amusement any attempt — be it in print, in the drawing-room, or at Exeter Hall — at overpiousness.

Charlotte Murchsion, in a long gossipy letter about recent doings in the London scientific set, also urged her to 'come over for the Cambridge party . . . [for your] absence . . . [would be] a great disappointment to them all'.[154] Mrs. Somerville, however — already quietly planning a summer jaunt to Switzerland — declined. Dr. Somerville attended with his old friend William Sotheby the poet. Sotheby celebrated the occasion with a long commemorative poem in the classical style.[155] In it he hailed the power of British science, the glory of Britain and of Cambridge, and the achievements of a number of British philosophers including some living ones.* Its penultimate stanza gently chides Mary Somerville, the only woman mentioned, for her failure to appear among her scientific peers:

> But — thou, in whom we love alike to trace,
> The force of reason, and each female grace,
> Why wert thou absent? Thou, whose cultured mind,
> Smoothing the path of knowledge to mankind,
> Adorn'st thy page deep stored with thought profound
> With many a flowret cull'd from classic ground;
> While Cambridge glorying in her Newton's fame,
> Records with his, thy woman's honoured name,
> High gifted SOMERVILLE? —[156]

Arrangements for publishing Mary Somerville's forthcoming work occupied her attention and that of her husband for a good part of the spring and summer of 1833. The manuscript was to go first to Murray but the couple were not at all certain that he would wish to underwrite such a work. The letters that passed between Paris and London in these months give a running account of negotiations from the Somerville point of view.

At the end of February Murray reported to Dr. Somerville that all but 30 of the 750 printed copies of the *Mechanism of the Heavens* were sold. The news gave as much pleasure in Paris as it did in London. Both young Mary[157] and Martha[158] wrote their brother exultantly of this success, complained that they had heard nothing from Murray about the new book, hoped that he would take the new one and that it would sell equally well. Later in the month even more substantial reward arrived from the publisher. He sent Dr. Somerville an 'account of the sales of Mrs. Somerville's Work, the profit of which coming to her is £142.17.6'.[159] As long as Dr. Somerville was alive, Murray's accounts and statements about Mrs. Somerville's books were made out to and sent to her husband.

This first statement showed total costs of £518.15.9 and reflected a steady rise in sales over the months since publication. A bill to William Somerville bearing the same date[160] listed charges for four special bindings: one for ten shillings, marked 'russia' which probably went to the Emperor of Russia, another £2.12.6, 'Green Morrocco [sic] Elegant for the Queen' (likely Queen Adelaide), and two in 'blue Morrocco' at eighteen shilling each. These costs reduced Mrs. Somerville's royalties by £4.18.6 and Murray in his letter made

---

*   Airy, Babbage, Brewster, Brisbane, Buckland, Conybeare, Compton (later Lord Northampton), Dalton, Davy, Faraday, Gilbert, Greenough, Herschel, Lindley, Lubbock, Lyell, Murchison, Peacock, T. R. Robinson, Roget, and Mrs. Somerville.

the sage observation that her profits would have been £222.17.6 had not '70 cops been presented in various ways' and, further, that if he had 'ventured to print 1,000 instead of 750, there would have been no profit at all — as there would have been 275 left in hand and a greater expence'. In a splendidly generous gesture, the publisher added,

. . . I do not intend to diminish this profit by taking any portion of it myself — for I am overpaid by the honour of being the publisher of the work of so extraordinary a person & have therefor charged only a commissionon the actual long paid expenses for Interest and bad debts.[161]

On learning of this action, Mary Somerville wrote her husband that Murray had 'behaved very handsomely indeed' and enclosed a note of thanks to him, asking Somerville to pass it on if he judged it suitable, which he did. In it she remarked:

When I consider the very unpromising nature of my work, and the small probability there was of success, I am more and more sensible of what I owe to your kindness and liberality . . . I am quite surprised at the number of copies that have been sold, and I must add very much pleased, for independently of myself, I should have been truly grieved had you been a loser by having generously undertaken what appeared so hopeless. I am happy to find you have ventured on my new attempt and trust it will be more popular. I have done all I can to make it so, and shall be glad of any advice you may give on the subject. . . .[162]

This deserved praise differed from some of the comments about the publisher that had hitherto found their way into letters from Paris. In February young Mary had written her brother:

I hope Papa will squeeze Murray well [in the matter of the new book]. I am sure Mamma deserves a good deal more than that horrid wretch Mrs. Trollope who got 1000 for her book! and poor M gets (praise of course as she deserves) but nothing substantial nasty Murray.[163]

The two girls seem to have found in Mrs. Frances Trollope (1780–1863) a rival — and one unjustifiably better rewarded — to their own mother, under-appreciated, in their eyes, by the publishing world. Two months earlier young Martha Somerville had commented to her brother[164] that Mrs. Trollope's new work was vastly inferior to her *Domestic Manners of the Americans,* published in 1832 and of interest to the Somervilles as they made American friends in Paris.

The report Murray gave of sales of the *Mechanism* excited all the Somer-villes,[165, 166] and his March magnanimity understandably altered to some degree the tone of their comments. The condescending attitude of the young ladies grew no doubt out of remarks heard at home and interpreted by inexperienced minds, but one recalls that Lyell — according to his biographer Wilson[167] — regarded Murray as more a successful tradesman than a gentleman. Whether the Somervilles' Scottish pride of birth fostered the imperious tones of these March 1833 letters — an explanation diminished by S. C. Somerville's long friendship with the Murrays — or whether they arose out of sheer callowness, they rapidly vanished. The association between the Somervilles and Murrays continued with utmost cordiality for half a century.

The unexpectedly good sales of the *Mechanism of the Heavens* must have been a factor in Murray's decision to publish Mrs. Somerville's second book. William Somerville apparently opened negotiations when he returned to London in late November 1832, for almost immediately young Martha wrote of her mother's eagerness to know 'if anything is settled with Murray' and asked, 'Would it not be a good plan to give this new work to Lardner's Encyclopedia?'[168] Dionysius Lardner (1793–1854; F.R.S., 1828) had just republished Sir John Herschel's *Preliminary Discourse*[169] in his Cabinet Cyclopedia series and was scheduled to bring out Herschel's new book on astronomy as well as Charles Dupin's new book on the useful arts, thus drawing their attention to the series.

The manuscript Murray received in the new year now incorporated the newest findings in physical science from both France and Britain. Mrs. Somerville, commenting on an amusing report of her son's 'jollification with John', added:

. . . I hope it will have a good effect, but Dr. S. wrote to me that the bibliopole had carried off the M.S. [in W.G.'s hand here, '** of the Connexion of the Physical Sciences'] to try if it was intelligible to himself before he could give an answer, a measure which I confess is not a little alarming as I dont believe he will understand a word of it. You shall hear as soon as anything is fixed.[170]

Greig much later inserted in the letter the identification '**the late John Murray of Albemarle St. the leading Publisher of London and King of Good Fellows. W.G.'.

From late February until well into March Mrs. Somerville awaited Murray's decision.[171, 172] Before the end of April he had the book in press,[173] and by the end of May proof sheets began to arrive for her inspection.[174] 'We all approve the size and type of the book', Martha Somerville wrote her brother, 'I am glad the lines are not closer. It is much more encouraging to reading when the lines are widish apart than when they are close'.[175] The printed book would be, in fact, about two-thirds as wide and as long as the *Mechanism*.

Through the kindness of Lord Granville, the sheets were sent back and forth, along with the Somervilles' personal correspondence, in the Embassy postbag;[176] this not only speeded delivery but saved considerable money.[177] Mary Somerville corrected proof through May and June, complaining nevertheless at the slowness of transmission[178] and unhappy that the book would not be out in time for the coming London publishing season.[179] She was anxious lest Sir John Herschel's new work on popular astronomy[180] cut into possible sales but concluded that her own work would have to 'take its chance'.[181] William Somerville in London was endlessly helpful, doing errands, transmitting messages and advising her on various points. Brougham began to show an interest in the new work and Somerville reported his suggestion that it be dedicated to Queen Adelaide. On 2 April Mary Somerville wrote her husband, expressing gratification at the Lord Chancellor's interest and adding, '. . . as to the dedication, do as seems best. What to say to a Queen I know not, you must tell me'.[182] She sought William Somerville's opinion as to whether her paragraphs were too long[183] and approved of his plan to have Brougham and Whewell correct the sheets as they came from the press. She would then look over the whole before the book was finally printed. No query was too trivial, no mission too delicate for Dr. Somerville to undertake. The correspondence in these months reflects utter confidence on her part in his judgement and full support on his for her efforts.

By May plans for a trip to Switzerland and perhaps Italy — the final touch to the young ladies' European education — were settled, the party made up — Dr. Somerville and Woronzow Greig were to join the ladies — and the itinerary fixed. Their departure, however, was delayed until early August. In June first young Mary, then Martha, became ill with the grippe, an epidemic of which was sweeping Paris as a month earlier it had swept London.[184] Then difficulties arose over Dr. Somerville's leave from Chelsea — a revision of organization was taking place there in which Lord John Russell (1792–1878; F.R.S., 1847) succeeded in abolishing various appointments but leaving intact the now dual office of Physician-Surgeon[185] held by his friend William Somerville.[186] Late in July the doctor finally managed a few days in Paris. He and his family went again to Lagrange 'to express . . . [their] gratitude and take leave of the Lafayettes'.[187] Plans were made for their return there after the trip to Switzerland.

But plans, during the next two months, did little but change. The Somerville ladies got themselves to Geneva but had to wait there for an extra week for Greig, delayed by the heavy traffic south from Paris. During this week Mrs. Somerville renewed acquaintance with the Genevese savants and their families and had opportunity to hear of current science in that country. When the Swiss tour was first broached, young Martha wrote her brother to urge 'putting Geneva at the end' on grounds that it would be difficult to get away from the city, that 'the Marcets, Franks, de Candolles, Sisimondis & all the learned clamjamfry . . . would . . . detain' them.[188] The young Somervilles were not always deeply respectful among themselves of their parents' scientific friends.

Greig and the rest of the party — his Cambridge friend William Smith O'Brien (1803–1864), (later a fiery leader of the Irish movement), his bride and his older sister — finally reached Geneva and for the next six weeks the group toured the Alps. Mary Somerville by this time could boast that she thought 'nothing of walking eight or ten hours and find my head as steady on a height as it used to be'.[189] Her health appeared to be completely restored. At the hospice on the Grimsel, 'one of the most out of the way, wild and desolate places' Woronzow Greig had ever seen, they were surprised and 'delighted . . . [to meet John Murray] with his friend Mr. [William] Brockedon [the artist (1787–1854; F.R.S., 1834)].' Murray gave them 'a good account' of William Somerville in London and told of his 'proposed plan of accompanying Mr. Murray's family to Paris', then escorting his own wife and daughters back to London, a 'scheme . . . [that would] enable the ladies to remain some little time longer in Paris . . .'.[190] Travel for ladies was incessantly complicated by the necessity of having a male escort. There had been a good deal of correspondence about the safety and propriety of unaccompanied ladies journeying from Paris to Geneva; in the case of crossing the Channel there was no question in the Somervilles' minds that a gentleman escort was requisite.

Whoever their attendant — and the details of their journey home are not revealed — the Somerville ladies were back in London for the 1833–34 season. All had profited from their months abroad. The girls, their mother wrote her husband, 'are now fitted to take their part in society with the first rank in any country'.[191] She had observed with pleasure the steady refining of her daughters' manners, tastes and style, the widening and deepening of their education and interests. As for herself, modest though she was, Mary Somerville had enjoyed to the utmost the celebrity the French accorded her. She had renewed acquaintance with the leading French savants and from them and others had learned at first hand of some of the latest findings in French science, material which she had incorporated into the manuscript of a second book, now completed. She had enjoyed pleasant and instructive occasions, had witnessed historic scenes and met celebrated persons, and, above all, had formed friendships she would always treasure. She returned to London a strong link in the bridge between the French and English scientific communities.

CHAPTER 7

ON THE CONNEXION OF THE PHYSICAL SCIENCES

1. The Physical Sciences, 1830–1833

When Mary Somerville began her second book her own ideas of what properly constituted 'the physical sciences' were clear, but a definition of the term was only then being formulated by scientific usage. The decisions she made about material to be included under that rubric were influential in characterizing it and were unchallenged. Today a sentence from her work is cited in the *Oxford English Dictionary,* under the second meaning of the word 'physics', as an example of 1834 usage. Her view of the compass of physical science was a modern one in the early 1830s. She excluded from consideration chemistry and biology and directed her attention to a 'group of sciences, treating of the properties of matter and energy . . .'.[1] Her plan is nowhere explicitly stated, but the book itself and a sentence from her memoirs confirms it: 'There were many subjects which which I was only partially acquainted, and others of which I had no previous knowledge, but which required to be carefully investigated, so I had to consult a variety of authors, British and foreign'.[2] The novelty of her work gave her a latitude of choice that future authors rarely enjoyed.

Spectacular successes in scientific investigations over the past half century had multiplied the number of sub-branches of science, as both old and new scientific intelligence was organized and re-organized into units small enough to be mastered and be of optimal usefulness. Increasing professionalization of science also called for more clearly delineated areas of specialization in order to give direction to the education of new professionals and to sub-divide conveniently scientific questions, scientific labours, even scientific jobs into harmonious and readily recognized categories.

Mrs. Somerville's work came at a time when there was need for new groupings in science. An examination of the most authoritative readings available to her from the turn of the century not only reveals the point from which she started but discloses a confusion that goes far to account for the readiness with which her views, once formulated, were accepted. Four eminent philosophers — John Robison, Thomas Young, John Playfair and John Herschel — successively touched on the classification of science in their writings prior to 1831.

Robison (1773–1805) in his piece on 'Physics' in the third edition of the *Encyclopaedia Britannica*[3] differentiated between *physics,* which he defined 'as being with us [the British] the study of the material system, including both natural history [the study of animate nature] and [natural] philosophy [the study of inanimate nature]', and what he called *general physics.* The latter encompassed, according to him, three branches: natural or mechanical philosophy (which he in turn reserved for the sciences of motion such as astronomy, hydrodynamics and optics), chemistry and physiology. His entry remained virtually unchanged through the seventh (1830–1842) edition of the encylopedia and was merely shortened drastically for the eighth (1853–1860).

Shortly afterward Thomas Young offered a more elaborate classification. In 1807 he published a revised and enlarged set of the lectures he had given in 1801–1803, as professor of natural philosophy at the Royal Institution. Mary Somerville knew his two volumes well and characterized them as 'a mine of riches to me'.[4] From them she became familiar with Young's categorization of the sciences, one of the earliest of the century. It was a murky one and in her book she avoided any slavish adherence to it. Young divided natural philosophy into three sectors: (i) mechanics; (ii) hydrodynamics (which subsumed hydraulics, acoustics and optics, all dealing with fluids); and (iii) physics, the last relating to 'the history of the particular phenomena of nature, and of the affectations of bodies actually existing in the universe'.[5] Under physics he placed, in a first instance, three sub-groups: (i) observational astronomy, geography and the properties of particles; (ii) heat, electricity and magnetism, and (iii) meteorology and natural history. He may well have wished to include chemistry in his overall classification of natural philosophy but in 1801 the professorship of natural philosophy and chemistry at the Royal Institution had been divided — Young becoming professor of natural philosophy and Humphry Davy lecturer (later professor) of chemistry — so Young omitted it. In Part III of his work, 'Of the Sub-divisions of Physics into Distinct Branches', he muddled the issue by listing five sub-groups: (i) statics, dynamics, pneumatics, hydrostatics, stereostatics, and crystallography; (ii) sound and light; (iii) astronomy and celestial mechanics, and geology; (iv) mineralogy and chemistry; and (v) heat, electricity and magnetism, plus a small amount of zoology and botany that fitted into his primitive underlying scheme. Not for several decades would Young's usage of the word 'energy' in Lecture VII of his *Course* to denote the quantity of work which a material system can do be recognized as fundamental in the definition of *physics*.

Professor John Playfair, one of Mary Somerville's early mentors in Edinburgh, sought as early as 1812, when the first volume of his *Outlines of Natural Philosophy* appeared, 'to have the elementary truths of Natural Philosophy brought into a small compass, and . . . arranged in the order of their dependence on one another'.[6] Before his death he had completed two of his three projected volumes and his plans for the third are known. His approach was in the tradition of Scottish mechanical philosophy, which conceptually assigned dynamics a central role and thereby ensured a more specific meaning to 'natural philosophy' than its usual sense of the general study of nature. Playfair, using as his guide a statement that, 'In the material world, the action which takes place among bodies, either produces a permanent change in the internal constitution of these bodies, or it does not',[7] allotted phenomena of the first kind to chemistry, those of the second to natural philosophy, while restricting natural history to collection and classification of facts. Under natural philosophy he grouped dynamics ('the most elementary branch of the doctrine of motion'[8]), mechanics (a modification of the doctrine of motion) — statics being a branch of these two — hydrodynamics (which in turn had four parts: hydrostatics, hydraulics, aerostatics and pneumatics), astronomy, optics, electricity and magnetism. Whewell in 1819[9] challenged Playfair's view that statics and dynamics were not distinct sciences and by 1833 had published five other works on the subject.[10] Mrs. Somerville knew of this opposing view, for in June 1832 she thanked him for his present to her of a copy of his 'valuable work on Dynamics'.[11]

It was Playfair's larger views rather than his sub-division of natural philosophy that influenced Mrs. Somerville in fashioning her second book. His ideas about the dependence upon each other of the truths of natural philosophy and his conviction — widely shared — that gravitation was the principle 'which pervades all Nature, and connects together the most distant regions of space, as well as the most remote periods of duration'[12] imbue her work. Playfair thought it probable that a similar principle obtained for non-gravitational phenomena (he cites heat, electricity, chemical affinity, cohesion, elasticity, impulse and galvanism) and was neither sanguine nor despairing of the discovery of 'some universal, unifying principle, operative right across the variety of natural phenomena, and conforming to the constraints of Scottish natural philosophy'.[13]

John Herschel, to whom Mary Somerville increasingly turned for guidance from 1830, approached the division of the field of science obliquely in his *Preliminary Discourse on the Study of Natural Philosophy*. In this influential book Herschel considered only inanimate matter. At a time when the pair 'natural science and natural philosophy' were parallels in the material realm to 'moral science and moral philosophy' in the spiritual, he employed the terms 'Natural philosophy', 'physical science', and 'physics' interchangeably. He brought one of the most informed scientific minds of the day to a consideration of the origin of the new divisions of science, concentrating on philosophical rather than cultural aspects of the matter. In 1830 he declared:

In pursuing the analysis of any phenomenon, the moment we find ourselves stopped by one of which we perceive no analysis, and which, therefore, we are forced to refer (at least provisionally) to the class of ultimate facts, and to regard as elementary, the study of the phenomenon and its laws becomes a separate branch of science.[14]

Keeping his criterion in mind, he used force, motion and matter as the bases of his categorization of the sub-branches of science; the resulting system has traces of the older classifications as well as evidence of the newer studies. Judging on the basis of cause and effect, Herschel saw

. . . two great divisions of the science of force, *Statics* and *Dynamics* . . . [each of which] again branches out into distinct subdivisions, according as we consider the equilibrium or motion of matter in the three distinct states in which it is presented to us in nature . . . The principles both of the statical and dynamical divisions of nature have been definitely fixed by Newton, on a basis of sound induction . . . [and] are competent . . . to the solution of every problem that can occur in the deductive processes, by which phenomena are to be explained or effects calculated.[15]

In Herschel's classification, the sub-divisions of physical science were (i) mechanical sciences (pneumatics, hydrostatics, and stereostatics), (ii) crystallography, (iii) acoustics, (iv) light and vision, (v) astronomy and celestial mechanics, (vi) geology, (vii) mineralogy, (viii) chemistry, (ix) heat, (x) magnetism and electricity, all listed in the order given. Zoology and botany were included insofar as they deal with fossil (and therefore inanimate) materials; these two sciences were his final 'distinct branches' of physics.[16]

Herschel's view of what properly constituted *the physical sciences* profoundly influenced Mary Somerville's thinking on the matter, as did his assertion of the 'mutual relation and dependence . . . between great branches of science'.[17] In preparing her 'Preliminary Dissertation', she wrote, she too 'saw such mutual dependence and connection in many branches of science' that she carried 'the subject . . . to a greater extent' in a second book.[18] To do so comprehensively, however, she was forced — as she pointed out in her memoirs — to familiarize herself with a number of sciences new to her.

Chemistry, long recognized as a branch of science separate from natural philosophy, was — to judge by Thomas Thomson (1773–1853; F.R.S., 1811) — equally slow in defining its sub-groups. From its first edition in 1802 Thomson's *System of Chemistry* incorporated the most advanced ideas on the subject, but not until its seventh edition[19] in 1831 — shortly before the appearance of Mrs. Somerville's new book — did he, for the first time, divide chemistry into four sub-groups: (i) heat and electricity; (ii) mineralogy and geology; (iii) inorganic chemistry; and (iv) organic chemistry. At the same time he urged, in his preface to the section on heat, that new chairs be instituted at universities to teach the principles of heat, light, electricity and magnetism, a combination he now judged mature enough to support a separate existence. These subjects were claimed also by the natural philosophers, and Mary Somerville included them in her 'physical sciences'.

Despite Thomson's inclusion of geology as a part of chemistry in his 1831 *System,* geologists in Britain had been publicly asserting the independent existence of this science since the founding of the Geological Society in 1807. Their case had been strengthened — as earlier the case of chemistry had been — by the establishment of a readership in geology at Oxford in 1813 (to which William Buckland was appointed) and the designation by Cambridge in 1818 of its Woodwardian professorship as reserved for geology (to which Adam Sedgwick was named). Mrs. Somerville, on close terms with these two philosophers and with many other geologists, made no move to include that science in her grouping. By 1830 several other sub-branches in the study of inanimate nature had also well established separate existences — astronomy, meteorology and mineralogy to name but three — while new designations, such as acoustics and crystallography, were familiar. These she did not hesitate to incorporate as 'physical sciences'.

She was aware also of the fact that the French since late in the eighteenth century had recognized '*la physique* . . . as a branch of science separate from mathematics on the one hand and chemistry on the other'.[20] The study of the general properties of matter, of heat, light, electricity, magnetism, and meteorology constituted 'la physique' which was taught by a 'professeur de physique'. Gay-Lussac did so and from him and other friends among the Paris savants she was familiar with this grouping. In Paris in 1832–33 she was busy with the manuscript of her book and easily susceptible to French suggestions.

The founding of the new British Association for the Advancement of Science and the categories it defined in setting up its organizational structure came too late to be of influence, even had their authority been more firmly established. Vernon Harcourt (1789–1871; F.R.S., 1824), the major figure in managing the first meeting, at York in September 1831, pointed out the dangers of detaching, one by one, single sciences from the 'central body . . . [lest] by degrees the commonwealth of science be dissolved',[21] but neverthe-less recommended that sub-committees be formed to deal with 'the particular advancement of every science'.[22] The divisions set out there and at subse-quent meetings over the early years of the organization are of interest in demonstrating the categories which were at the time useful and attractive to philosophers, while the changes in number and scope of these sections reflect not only administrative convenience and attempts to strengthen the new body but responses to developing needs in science. Mary Somerville in successive editions of her book would show the same consonance with scientific prog-ress. She was always aware, through her many friends among the leadership of the B.A.A.S., of decisions, discussions, and happenings in the group.

In all likelihood her selection, over 1831 to 1833, of the sub-branches of science to be subsumed under 'physical sciences' posed no problem to her. Nowhere in her papers is there any hint of difficulty in making her choices or any request for advice. Nor was she given to pondering philosophical niceties. She had at her command, as she set about this task, a familiarity with the best recent writings on the subject, a large array of the most recent scientific information, and a vast stock of common sense. For years she had heard science discussed by the most distinguished philosophers at home and abroad and was aware of their opinions on most scientific matters. She could call upon them freely for help and criticism. She was keenly aware of the state of each science, the degree to which it was considered by competent practition-ers formed and independent. She had lived in France and knew the French concept of 'la physique'. She simply drew upon all this knowledge and experience and set out what she thought constituted 'the physical sciences' and in doing so formulated a received opinion.

An analysis of the contents of *On the Connexion of the Physical Sciences* reveals that those sciences with which she was most familiar receive the greatest coverage, those with which she was less well acquainted the least. Further, her choice of sub-branches of science emphasizes the modernity of the work, for she includes topics most attractive to up-to-date scientific practitioners in the 1830s — astronomy, meteorology, electricity and magnetism (including terrestrial magnetism), light (with much attention to polarization), and optics, even tidal observations. Her extensive treatment of electricity and magnetism, second only to that of physical astronomy, is unusual for general works of this date. The last sections of her book on comets, stars, nebulae and other celestial phenomena of this sort — sections far removed from the opening ones on physical astronomy — hint at the interest already excited by the expected reappearance in 1835 of Halley's Comet; in the first edition these sections have every appearance of a last-minute addition but this order was maintained through subsequent printings.

The passion for quantification which her 1817 letters from France had reflected is evident also in this text. Numbers are frequently cited, and there are occasional references to the existence of a mathematical formula for some specified scientific purpose but the formula itself is never given. This absence of mathematics rendered the book, to her mind, 'popular'. This designation, however, did not imply any lowering of her text. Readers might be unversed in mathematics but they were required to bring a high level of attention and a cultivated vocabulary to perusal of the book. Nonetheless Mrs. Somerville never wrote condescendingly and always with admirable clarity and style.

The text is divided into sections, rather than chapters, and their lengths vary. In her first edition Mrs. Somerville covered the following general subjects — the 'physical sciences' whose connection she explored — in the order and to the extent given (see Table I).

Table 1. Organization of *On the Connexion of the Physical Sciences*

Subject	Number of Sections	Number of Pages
Physical Astronomy	14	123
Matter (atoms and molecules and the forces that hold bodies together or separate them)	2	14
Sound	2	25
Light and its Phenomena, including Vision	6	65
Heat (and temperature, meteorology and physical geography)	3	57
Electricity and Magnetism	8	74
Descriptive Astronomy (comets, stars, nebulae)	2	47
Conclusion	1	7
Explanation of Terms	—	36
Index	—	6

The basic pattern would obtain through the ten editions of her work. Discredited material dropped out, new material came in, and there were sometimes minor rearrangements — but Mrs. Somerville's classification of the sciences remained essentially the same over nearly half a century.

At a time when categorizations based on such general principles as 'statics', 'dynamics', 'hydraulics', 'pneumatics' and the like were widely popular Mary Somerville does not enter into such divisions. She is aware of branches of mechanics and on occasion, in dealing with certain points, refers to such terms. Her chief interest centers always on phenomena and the explanations for them. Her search is for connections between the physical sciences, not for categories. The fact that she in this work formulated an unchallenged definition of physical sciences is incidental to her main purpose.

Scientific readers — even the most expert — accepted her view that the physical sciences encompassed astronomy (both physical and descriptive), meteorology, physical optics, heat, sound, physical geography (and a bit of incidental geology, always of the latest sort), mineralogy (and some crystallography), electricity and magnetism, and a consideration of constituent bodies and their behaviour under various conditions. They made no criticism of her choice nor were there charges of omission of any large subject. As for general readers, Mary Somerville's book formed their notions of what constituted the physical sciences and did so for almost three generations.

2. The Final Revision

In the weeks following her return to Chelsea, Mary Somerville not only resumed management of family and household affairs and her customary London social life but revised her forthcoming volume in the light of comments made on her text by a number of distinguished consultants. Letters in the Somerville Collection indicate how expert and how detailed was the advice she received. Her printed text reveals how closely she followed it.

The general plan of the book was hers and seems not to have been altered significantly in the last three months of 1833. She had settled on which sciences and which connections she wished to present and had gathered an immense quantity of recent scientific material. Her revisions, with one exception, were of details — correction of numbers, clarification of passages, elimination and addition of material, polishing of text. Their total extent is unknown, for no complete manuscript of the work is found in the Somerville papers. What does survive are sheets in her hand, clearly related to sections of the book but undated and fragmentary, and a number of letters from her consultants. Some of these missives reflect painstaking throroughness in reading her sheets. For the most part they deal with minutiae; only occasionally do they touch on conflicting interpretations of phenomena or give insight into a writer's own views on controversial points. Their tone is uniformly approving. The approbation Mary Somerville's book met with from her corps of consultants and the praise it received when published can be taken as a general endorsement both by scientists and by non-scientists of its contents and of its interpretations of natural phenomena.

In making use of the services of many of their scientific friends, the Somervilles were following a common practice. Experts expected to be asked for advice and information and to read manuscripts or sheets of forthcoming works in their own or related fields. They in turn sent their *own* creations to friends for the same sort of careful attention. Not only papers[23] but proof-sheets were often read by several authorities before actual printing. Charles Lyell, W. H. Fitton, and Adam Sedgwick, for example, each read Charles Babbage's *Ninth Bridgewater Treatise*[24] before that controversial work appeared.[25, 26] At a time when a professional apparatus for scientific publication was largely lacking, this system of informal refereeing acted to insure a high level in published scientific works.

Mrs. Somerville was fortunate in the number and distinction of her consultants. Letters in the Collection reveal that at least six persons —Brougham, Faraday, Forbes, Henry Holland, Lyell and Whewell — saw all or some of the sheets before the book appeared. John Herschel and Charles Babbage, who had been principal advisers in the preparation of her *Mechanism of the Heavens,* seem not to have taken any part in the new endeavour. Herschel was busy making preparations to take his family to South Africa to spend some years there. In July 1833 he had presented his catalogue of the nebulae visible in the northern hemisphere to the Royal Society. With that great review done and after the death of his aged mother, he proposed to make the same sort of survey of 'the celestial phenomena of the southern hemisphere'.[27] Before the Herschels left on 13 November, Lady Herschel wrote Mary Somerville regretting that they would not be able to have the 'long *chat* . . . respecting your own past travels & our projected wanderings, but my husband can find no time for any thing but business'[28] and assured her of their continued attachment to the Somervilles. Babbage, caught up as ever with his machine, felt his talents lay chiefly in mathematics and machinery, not the physical sciences. He and Mary Somerville, however, remained in close touch; in December 1833, for example, he wrote that he had a copy of Herschel's paper on nebulae for her.[29]

During her absence in Paris William Somerville not only made enquiries for her but dispatched sheets to Whewell and Brougham for preliminary inspection. Whewell, immediately after completing the 'employments . . . bequeathed' by the B.A.A.S. meeting in Cambridge and moving from one set of rooms to another in Trinity College, turned to a close perusal of Mrs. Somerville's work. Notes and letters from him to William Somerville, accompanying the returned sheets,[30] testify to the thoroughness with which he carried out his task. In late summer while still on holiday, for example, he sent from Liverpool ten different detailed comments and corrections on Mrs. Somerville's treatment of light (from pages 197 to 218 in the printed text), in each instance indicating the page and line.[31] On returning to Cambridge, he sent Dr. Somerville 'the vocabulary', some corrections and rewritings, some advice on the illustrative figures and on mineralogy.[32] Toward the end of November, some weeks after Mrs. Somerville's return from abroad, Whewell was still correcting sheets newly come from the press, still declaring that 'as usual [he had] been very free' in his remarks, but adding 'Mrs. Somerville can easily afford her critics an indulgence which other Ladies would not dream of'.[33]

Brougham was less diligent but did attend to some of the material. Since early spring he had concerned himself with the dedication of the work. On 12 September, writing Dr. Somerville to acknowledge receipt of 'sheets', he added, 'I ought long ago to have announced to you, Mrs. S., the Queens acceptance of the Dedication which H. M. has done in the *most gracious manner* possible . . .'.[34] After her return to Chelsea Mrs. Somerville prepared the text of the dedication, which was to be dated 1 January 1834, and sent it to Brougham. He returned it, along with comments qualifying for quotation since they indicate his unyielding nature and strong feelings about his old quarrel over the undulatory theory of light with Thomas Young, a philosopher now almost five years dead:

I enclose the dedication — & only doubt whether it should end with "humble" or "devoted" If you & the Dr. think the latter — then instead of "profound" — it should be "the greatest" —

Many thanks for the sheets — which I have read with equal pleasure & instruction as those I formerly had from you — One or two things I could have troubled you with — but they are of little moment. I shall note them — & give you them hereafter. The only one that is at all material relates to the way you mention Dr. Young — not that I object to "Illustrious" &c as applied to him. But as you dont give it to one considerably more so, it looks either as if you overrated him or underrated Davy — or (which I suppose to be the truth) as if you felt Y. had not had his due share of honour — & desired to make it up to his memory. Observe I give him a very high place — But Davy's discoveries are both of more unquestioned originality & more undoubtedly true — perhaps I should say more brought to a close — The alkalies & the principle of the Safety Lamp, are *concluded* & fixed — the undulation is in progress & somewhat uncertain as to how and where it may end — You will please to observe that I reckon both those capital discoveries of Davy the fruit of inquiry & not at all of Luck for as to the lamp, it is plain & as to the metals, if you look at the inquiries that immediately preceded, you will see he was thereby led to the alkalies — Indeed I well remember saying when I read them "He will analyze lime & barytes" —

I am quite ready to admit his *extreme* folly in some things — but that is nothing to the present purpose — & indeed I daresay Young had his full share — I have heard of his ridiculous dancing and writing Greek on a sixpence & indeed both Davy and he had the supreme folly of giving up their original and natural opinions for the love of Lords and Ladies.[35]

Brougham's opinion did not alter Mary Somerville's. She retained the adjective 'illustrious'[36] and continued to support Young's undulatory theory.

The copy of the dedication to the Queen found among the Somerville papers clearly delineates Brougham's contribution to this inscription. Written in Mary Somerville's hand and passed on to him through William Somerville, it carries the Lord Chancellor's emendations and suggestions in his own. At the top of the sheet is the query in Mrs. Somerville's writing, 'What ought I to begin with?' answered in Bougham's hand, 'To the Queen'.[37]

In addition to corrections and comments from Whewell and Brougham, a few come from J. D. Forbes, who had unexpectedly added himself to Mrs. Somerville's corps by a note written Dr. Somerville on 20 August:

When in Paris Mrs. Somerville mentioned to me that though her little volume was almost ready it would not be out till Christmas. As I am now writing my Introductory Lectures I should be very desirous if possible to see it, both to profit by its contents, & to be able to give some account of them. It occurs to me that it is just possible that you might be able to further this object & would be willing to do so. Should it be impracticable you will I am sure excuse me for troubling you with this note.[38]

In the important first term of his professorship, Forbes was intent on making as good a case for his election as possible. Presentation of some of the newest work in physical science would enhance the tone of his lectures and reflect credit on him. Queries to senior male colleagues at home and abroad might, at this stage, diminish him in their eyes; queries to Mrs. Somerville, equally effective, would add to his reputation for knowing immediately what was going on in the world of science.

Dr. Somerville was willing to help. On 21 October the young professor sent him, with apologies for his delay, thanks 'for the sheets of Mrs. Somerville's delightful work' and confessed he was waiting 'with much impatience for the remainder'. Forbes also enclosed four minor corrections. He regretfully refused William Somerville's expressed 'wish' that he review the work for the *Quarterly,* observing somewhat tartly, 'I should have thought it necessary to have a communication from Mr. Lockhart on the Subject' and pleading lack of time 'at present . . . to undertake even that agreeable employment'.[39]

Dr. Henry Holland's contributions to the work were more substantial. Always interested in recent developments in science, Holland (1788–1873; F.R.S., 1816) kept a handy record — useful to himself and his friends — of reports of scientific facts and findings, of instruments and equipment, of experiments and experimenters and their published works. Just when he became involved with Mrs. Somerville's researches for the book is unclear but by 26 October 1833, he appears to have been familiar with a good part of the work. Two letters from him in the Somerville Collection, neither precisely dated,[40] give references and information clearly connected with the sections on electricity and magnetism. In one he asks also 'to see the optical part of your treatise — that also which relates to crystallization . . .'.[41] A third letter discloses Holland's belief that he has 'gone through . . . all those portions [of the sheets] in which recent discovery or controversy are more especially concerned' and offers his assessment of the whole work:

It is wonderful to me how completely you have encompassed & brought within your field all the greatest attainments of modern science; & very especially those which concern the great forces, acting on Matter, Electricity in its several forms, Heat, Light & etc. You have clearly & justly expounded the correlation & convertibility of these forces. The sole deficiency I find in this part of the work is that you have not *sufficiently* dwelt on the chemical affinity as one of them; & on the marvellous attainments of modern Chemistry in the creation as well as the analysis, of the various series of organic compounds; & in all that concerns & is derived from the doctrine of definite proportions . . . Altogether, my dear Mrs. Somerville, I think your book a very wonderful one — almost too good for the average class of readers. . . .[42]

In this letter he also states that he has added a note to a piece just completed by him for the *Quarterly Review*[43] — he mistakenly writes 'Edinburgh' — 'speaking of this . . . new work, as it ought to be spoken of' and that Murray, to whom he had shown his comment, was 'much satisfied'.[44] The publisher was, in fact, so pleased that he used the praise[45] in various advertisements of the book[46] for a number of years.

Holland's footnote in his anonymous review of four Bridgewater treatises gave the impression that Mrs. Somerville's book had already appeared. Charles Lyell, coming to town in December, enquired for it at Murray's only to learn that prospective readers must still have 'patience for a time'.[47] He wrote Mrs. Somerville to ask if she would 'allow . . . [him] to get the sheets from Murray whenever he thinks it may be done with propriety'. Lyell was at the time preparing the third edition of his *Principles* and busy educating himself in other sciences, especially chemistry which he now considered basic to the development of geology.[48] He may have hoped to learn something of that science — he is unspecific on the points 'on which . . . [he] anticipated much edification' in reading her work,[49] but was intent on getting the volume as soon as possible. His letters and Forbes' earlier appeal demonstrate the felt need for a work such as hers at the time.

Sheets went to Lyell, with whom the Somervilles spent Christmas Day, and during January and February — indeed up to publication day — he sent along comments and asked for further references on points that interested him.[50] Two of his corrections were of mistakes already noted by Forbes. It was too late, however, to alter the text. A page of errata, inserted near the front of the book, had to serve, listing corrections on fourteen different pages of text. In the second edition these changes and others were incorporated into the printed text.

The most authoritative comments on Mrs. Somerville's work are found on four large sheets covered with Michael Faraday's neat handwriting.[51] They deal with her sections on electricity and magnetism (pages 285–359), and are meticulous, identifying passages by page and by line. In some instances Faraday explains why he advocates alterations in the text or why he raises certain points. Whatever the original text might have been, the printed text in every case conforms to Faraday's advice. On occasion his very words are adopted. The date of the sheets of 'remarks' which he sent to William Somerville for his wife adds significance to them, for it is November 1833.

Faraday published his first series of experimental researches in electricity in 1831, his second and third series in 1832, and devoted 1833 to carrying out the fourth (dated 15 April 1833), fifth (June 1833), sixth (30 November 1833) and seventh series (31 December 1833).[52] Mary Somerville's sheets seem to have arrived as he was nearing the end of his seventh series of researches, one that continued his work on electrochemical decomposition and resulted in his enunciation of the laws of electrolysis. He presented the findings of the sixth and seventh series to the Royal Society at meetings in January and February 1834, completing his discourse on 13 February. Mrs. Somerville's book appeared a fortnight later.

The first edition of *On the Connexion of the Physical Sciences* does not mention Faraday's laws of electrolysis, but these, with other items from his sixth and seventh series of experimental researches made their way into the second edition, dated 1835 but actually printed late in 1834. Thus, this second edition has one of the earliest — if not *the* earliest — popular accounts of Faraday's fundamental discovery of the relation between the quantity of electricity and the mass and equivalent weight of a substance undergoing electrolysis.[53]

Faraday and his work are cited seventeen times in the first edition of the *Connexion of the Sciences*. Only John Herschel, mentioned eighteen times, stands above him. Laplace is referred to sixteen times, Biot ten, and Arago nine. Most authorities are mentioned only once or twice. Mrs. Somerville's recognition of the value of Faraday's work at a fairly early point in his distinguished career is notable, as is her extensive coverage of the science of electricity, interpreted according to his views.

Faraday's first remark on his four sheets for Mrs. Somerville is directed to her initial paragraph on electricity. He proceeds through her sections — his remarks cover about four-fifths of her text on electricity and magnetism — designating in all 27 separate passages for comment. In five of these instances, her printed text repeats Faraday's own words. In two instances he goes beyond scientific fact, figures or interpretation of researches to give his own opinion of the soundness of conclusions or the manner in which they were reached. One such remark is especially interesting in the light of Faraday's interruption of his electrochemical studies in November 1833 —the time of these comments — for an investigation of intermolecular forces and the catalytic action of platinum. His observation is one typical of the careful attention he gave Mary Somerville's work. She had, apparently, accepted uncritically the thesis that electrical repulsions could be attributed to the mechanical pressure of the atmosphere. Faraday responded:

I think it very doubtful whether the attractions & repulsions can thus be referred directly and mechanically to the pressure of the air which is of course implied by pressure upon the air. I think Electricians have hurried the science here & given a clumsy mechanical view because they could not perceive a better one.[54]

In the published text Mrs. Somerville, in conformity with this comment, states that 'it is doubtful whether these phenomena can be referred to that cause [pressure of the air]'.[55]

Faraday's preference for the pictorial over the mathematical in his analysis of phenomena is a subject of recurring interest. One of his remarks on these sheets — evoked apparently by some reference to philosophers predicting phenomena through the use of mathematics — illuminates his own attitude toward mathematics in science:

'Predicted' I do not remember that Mathematics have *predicted much* Perhaps in Amperes theory one or at the most two independent facts. I am doubtful of two. Facts have preceeded the mathematics or where they have not the facts have remained unsuspected though the calculations were ready as in Electro magnetic reactions & Magneto electricity generally and sometimes when the fact was present as in Aragos phenomenon the calculations were insufficient to illustrate its true nature until the facts were in to help.[56]

In the section to which he refers, Mary Somerville's published text records that 'the analytical investigations of M. Poisson and Mr. Ivory' had successfully determined the distribution of electricity and 'that all the computed phenomena have been confirmed by observation'.[57]

At the end of his fourth sheet, Faraday observes, 'I just wrote down at the moment whatever occurred to me. Do not therefore think me too free but treat these remarks just as freely.'. Because Faraday's sheets show such thorough and careful reading of Mrs. Somerville's sections on his part and because her printed text reflects so faithfully the comments and suggestions the philosopher made, a justified conclusion is that her work represents Faraday's actual views at the time it was published. Subsequent revisions made on his advice maintained this state of affairs. It can be taken that changes in her text mirror changes in Faraday's views, that at any one time her book represents his opinions at that time.

Sometimes early in January the final revisions were done. Well over half the experimental work included had been done on the Continent, and Mrs. Somerville's treatment of recent French work was exceptionally full. Yet none of her consultants in this last revision of her text were French. Language may have been one problem, distance another. The book which appeared, however, brought to its English readers the newest science on both sides of the Channel and filled a void that was already recognized.

3. Publication and Review

Mary Somerville's *On the Connexion of the Physical Sciences* was published in late February 1834. John Murray printed 2000 copies, priced at 7s. 6d. each. Eleven of these went to Stationers Hall, eleven to 'Reviews &c.' and twenty-one to the author. His costs were £422.15.6, including £2 paid for a poor index and £42.3.6 for advertising.[58] The book was ninth in the list of seventeen publications advertised as 'nearly ready for publication by Mr. Murray' in the 15 February issue of the *Literary Gazette*.[59] It was described as

. . . Being the substance of the Essay prefixed to the "Mechanism of the Heavens" enlarged and adapted for the general and unscientific reader, in 1 small vol. similar to those of Herschel and Babbage.

In his next advertisement, on 8 March, Murray substituted Holland's foot-note in the *Quarterly*:

. . . "Mrs Somerville's delightful volume on the 'Connexion of the Sciences.' The style of this astonishing production is so clear and unaffected, and conveys with so much simplicity, so great a mass of profound knowledge, that it should be placed in the hands of every youth, the moment he has mastered the general rudiments of education." — *Quarterly Review.*[60]

An identical advertisement appeared in the *Athenaeum*[61] on 15 March, again in the *Literary Gazette* on 22[62] and 29[63] March, and in the *Athenaeum* of 3 May.[64] Murray placed them for the benefit of booksellers, for his supply of copies was already gone.

Sales of the new work were extremely brisk. By 6 March Murray had disposed of all but eight or ten copies. Mrs. Somerville, sending her son this good news, remarked that

. . . no second edition will be called for till the booksellers have disposed of theirs but I have no doubt it will be req^d in time. Murray gives credit to the trade so I cannot expect to *bag* any game till he receives his money.[65]

In June, however, Murray sent her royalties of £144.18.10 — two-thirds of the profits earned to date.[66] This generous arrangement — two-thirds of the profits to her, one-third kept by Murray — was to obtain throughout the rest of their association and was Murray's idea. His usual habit was to give half the profits to the author.

Furthermore, the books went quickly in the bookshops. The Reverend Joseph Romilly (1791–1864) of Trinity College, Cambridge, for example, was sent a copy by his bookseller along with an anonymous commonplace book called *The Doctor*[67] when he asked for reading to take along on his spring holiday.[68]

The instantaneous success of *On the Connexion of the Physical Sciences* was due in good measure to two favourable early reviews in widely read weeklies. On 8 March the *Literary Gazette*[69] 'cordially recommended . . . [the work] to the general perusal of all who wish to be agreeably initiated into a love of science', a very large group in Britain at the time. Mrs. Somerville's style and skill were praised. On the following Saturday, 15 March, the *Athenaeum* in a piece interspersed with many quoted passages and running slightly more than a full page, called the book

. . . a most delightful volume, — with the exception of Sir John Herschel's treatises, the most valuable and most pleasing work of science that has been published within the century . . . Her book is at the same time a fit companion for the philosopher in his study, and for the literary lady in her boudoir; both may read it with pleasure, both consult it with profit. . . .[70]

The reviewer — it may have been Augustus De Morgan — was clearly familiar with the material covered and differed slightly with Mrs. Somerville in the interpretation of 'Dr. Young's celebrated experiment of the tuning fork'. He called her 'sections on Sound . . . the best in the book' and recommended the ones on electricity 'for an entertaining as well as correct summary of the most recent achievements in the field of experimental science'. The publication of the volume he regarded, as many of his contemporaries did, 'as an honour to our age and country'.[71]

Among the weekly publications catering to the self-help movement, the *Mechanics Magazine* took early note of the new work. On Saturday, 29 March, it published a long and awestruck review declaring, in much the same vein as had the *Athenaeum,* that the treatise was 'as well-adapted for the youth of seventeen as for his gray-headed preceptor — for the humble but intelligent mechanic as for the refined and talented philosopher'. Mrs. Somerville was described as a

. . . highly gifted authoress . . . [occupying] a conspicuous place in the very foremost ranks of profound mathematicians . . . already well-known to all the learned; and by [this] noble effort . . . [certain to] become through all future time equally familiar to the unlearned of every nation in the civilized world. . . .

Declaring that the 'cultivation of scientific knowledge is at once the glory and the peculiar characteristic of the days in which we live', the reviewer endorsed 'the spread of knowledge' and the necessity 'to enforce and preserve the "connexion" of [the] several departments [of the physical sciences]'. Twelve long extracts from the book concluded the review; its final words are, 'We do not . . . advise its being placed on the shelf. Instead of that we say — read it! read it!'.[72]

Since the previous summer, William Somerville had busied himelf in the matter of reviewers for the *Edinburgh Review* and the *Quarterly Review.* He was anxious that the choices fall on competent and friendly persons who would have their pieces ready as soon as the book came out. For that reason he turned to those already familiar with Mrs. Somerville's sheets. Forbes, one of his early prospects, begged off from lack of time.[73] Mrs. Somerville herself asked Henry Holland but he too regretfully declined, pleading press of business and hinting at some coming change in his life.[74] He did, in fact, remarry early in 1834, his bride being Sydney Smith's daughter Saba. In refusing the author's request, however, he urged Whewell as the person best able

. . . from the variety of his physical knowledge, to do it [the book] justice — The Review could not be satisfactorily executed by any one whose knowledge was limited to one or two branches of science. It is the *general view of relations* present & prospective, which is most to be desired in one who is to treat of the value & design of your work.

The Somervilles next approached Lyell, but he declared he had neither 'leisure or knowledge' sufficient to accept Lockhart's invitation.[75] Finally Whewell, who was familiar already with the sheets, agreed to prepare an account for the *Quarterly*. His anonymous review, which appeared in the March number,[76] has several interesting aspects apart from the exhaustive account it gives of Mrs. Somerville's work. It reflects some of the currents that were flowing through British science at the time and the responses that rose with utter naturalness from the reform group in the scientific community.

From the first it is evident that the reviewer is a modern man of science, aware of changing needs and attitudes, supportive of new studies and familiar with old. Mrs. Somerville's book, indeed Mrs. Somerville herself and her mathematical and scientific pursuits, are accepted and admired as part of these desirable changes. The reviewer lauds her 'exposition of the present state of the leading branches of the physical sciences', calls her a 'person of real science' unlike many popularizers, and includes in his review a number of passages from the work, among them one on comets and several on electricity and magnetism, stressing in the latter cases 'the novelty of the subject' covered. Deploring the 'increasing proclivity to separation and dismemberment' in the sciences, Whewell praised Mrs. Somerville for illustrating the 'connexion', rather than the division, of the physical sciences and went on to suggest that a new word — not 'philosophers . . . [not] savans . . . [was needed to] designate the students of the knowledge of the material world collectively'. Commenting that the want of such a general term had been 'felt oppressively' by the members of the British Association for the Advancement of Science at their meetings at York, Oxford and Cambridge, he recommended a new term, 'scientist', pointing out that its formation was analogous with that of 'artist . . . sciolist, economist, and atheist'. The B.A.A.S. is said to bring together 'the cultivators of different departments' of science, Mrs. Somerville to have a similar object in terms of the physical sciences. It is in this review that Whewell's suggestion — widely adopted — was first made public and a new term added to English usage.

The final quarter of his review he devoted to a topic that Mary Somerville's contemporaries often pondered: the differences if any between male and female intellect. Some of her friends and admirers considered Mary Somerville unique; others believed that intellectual women, if they desired, were capable of mastering knowledge. She herself was convinced that the chief sexual difference was the absence of creativity in women, a preponderance in the female of what she termed 'earthiness'. Whewell at the time of writing was brooding over his history of the inductive sciences — hints of this are evident in the review — and offers his own speculations and his conclusion that there is 'a sex in minds'. In the woman the heart rules the head, although the female philosopher is both profound and clear. In the male 'practical instincts and theoretical views are perpetually disturbing and perplexing each other'.

He concludes his review with two unattributed bits of Cambridge versifying. The first is his own sonnet to Mary Somerville, the second is modelled on Dryden's 'Under Mr. Milton's Picture', with Hypatia, Maria Agnesi and

Mary Somerville replacing Homer, Virgil and Milton. Its final four lines suggest some of the reasons Mrs. Somerville was so highly prized by her contemporaries:

> Equal to these, the third, and happier far,
> Cheerful though wise, though learned, popular,
> Liked by the many, valued by the few,
> Instructs the world, yet dubbed by none a Blue.[77]

The account of Mrs. Somerville's book in the *Edinburgh Review* while equally laudatory took a different tack. William Somerville had written the editor, Macvey Napier, sometime in November 1833 about the matter and Napier promptly replied that he would do all in his power 'to have the work properly noticed'.[78] Ten days later he sent the news that Sir David Brewster had agreed 'to write an article, for the April No.'.[79] After the New Year Brewster himself asked Mrs. Somerville to send sheets to him 'if you are desirous of having the Review inserted in the April number'. He also enclosed copies of two of his forthcoming 'optical papers'.[80]

To a discerning contemporary reader, the identities of the two anonymous reviewers should have been evident from the emphases and tones of their pieces. Whewell's review has the nonchalance of an already successful philosopher, securely ensconced in the scientific establishment and sure to rise higher, one easily familiar with its cleverest minds and its most influential figures. His account demonstrates a mastery of a wide sweep of learning, knowledge of the latest science, an interest in linguistic invention, a flexible and enquiring intellect sympathetic to change, capable of lightheartedness. In contrast Brewster is always serious, incessantly rendering judgements, and given to a display of his own expertise. He was an unhappy man in 1834 and his review subtly reflects his insecurity and forlornness. Still galled by his failure to win the Edinburgh chair, without a permanent situation suitable to his considerable talents, he was forced to divide his efforts between scientific investigations and editing and writing. Impatient with the old attitudes toward science and scientists, eager for change, he found himself still scrambling for livelihood in Edinburgh, far removed from the centre of reform.

Brewster's account,[81] more narrowly focused and critical than Whewell's, emphasizes those parts of Mrs. Somerville's book related to his own specialities. Almost half his essay is devoted to her treatment of light and vision, although it fills but 15% of her total text. He refers approvingly —twice by name — to his own recent work, remarks on her 'decided undulationist views' and makes some suggestions about specified statements and about rearrangements of material. The absence of any mention of the work of Eilhard Mitscherlitsch (1794–1863; F.R.S., 1828) and the want of 'ocular delineation . . . in imitation of Laplace' are deplored. On the whole, however, he finds the work 'a condensed and perspicuous view of the general principles and leading facts of physical science, embracing almost all the modern discoveries which have not yet found their way into our elementary works'.

At the end Brewster expresses reluctance to press Mrs. Somerville, able though she was, for further treatises; his reasons suggest his own situation:

. . . Mrs. Somerville's great mathematical acquirements, her correct and profound knowledge of the principles of physical science, and the talent for original enquiry which she has already evinced in her paper on the magnetism of the violet rays, induce us to urge her to original investigation . . . The fame of scientific authorship is but a poor compensation for its toils; and the fleeting celebrity of writing the best book upon a science which is undergoing continual change, and demanding new expositors, cannot gratify a mind like hers.

Both Brewster and Whewell single out for quotation the passage in the book dealing with the close relation and possible influence of light, motion and electricity on crystallization. Whewell approves, Brewster does not. What the latter could not know — but what is clear from the Somerville correspondence — is that Faraday in his 'remarks' of November 1833 had commented on the material and had not questioned the conjecture which Brewster now labeled dubious. Nor could Brewster have known that Whewell himself in August 1833 had read and generally approved of Mrs. Somerville's treatment of light; the nine suggestions which he did make were all incorporated into her text. Two of these revised passages were among those that Brewster specifically disparaged.

The *British Critic*, organ of the High Church party, was the third quarterly to notice Mrs. Somerville's new book. Its review was directed toward the 'Character of Modern Works on Physical Science' and stressed the connection between pysical science and 'revealed Religion'.[82] The reviewer deplores the paucity of 'allusions to the connexion between science and our moral and religious feelings . . . [found] in Mrs. Somerville's treatise' and praises Herschel's *Preliminary Discourse* and Whewell's Bridgewater Treatise. Mrs. Somerville is termed 'the first woman of her age' but her new book is declared inferior to her first and not recommended.

Of the three reviewers, it is the one from the *British Critic* who gives, in his closing words, the most balanced assessment of the value of Mrs. Somerville's work to the ordinary reader:

. . . As a book of reference to the most lately ascertained conclusions it may be useful; but certainly to the beginner in science it would be perfectly unintelligible, and as digest for one more advanced it is incomplete.

The quick sale of Mrs. Somerville's book and the praise heaped upon it in the press confirm the popularity of science among literate Britons in 1834 and their eagerness to have an authoritative account of the latest findings in physical science. Her book appealed to many on its own merits, to others because it enabled them to appear fashionably informed and advanced. The reviews, in stressing how superior and current was this work by a British woman, bowed to national pride. Argument over the alleged decline of science was no longer a burning issue in the press. Mary Somerville's second book appeared at a relatively comfortable public juncture in British science and added to the sense of national well-being. Her reputation assured the work prompt notice; its merits, ready sales. Reviews in the *Quarterly* and the *Edinburgh* not only gave, in the first case, a new term to English usage but revealed differences in style and attitude between practitioners at the centre of British science and those more remote from it.

The notice which the *Connexion* received in the English press brought the work quickly to the attention of American publishers. Key and Biddle in Philadelphia issued a pirated American edition[83] before the year was out. Professor Benjamin Silliman of Yale (1779–1864) mentioned the 'republication' in the 1835 issue of his *American Journal of Science*.[84] By 1837, 'copious extracts' from this edition were being used to inform possible investors in an 'Electro-Magnetic Association' of what the promoter termed 'the comparatively modern sciences of electricity and magnetism'. A brochure[85] issued in New York and describing a rotatory magnetic device invented by a Vermont blacksmith devoted more than half its text — some 60 out of its 94 pages —to reprinting portions of Mrs. Somerville's sections on electricity and magnetism and explanations of terms. The machine was exhibited in New York and came to the notice of the English public through the *Literary Gazette*.[86]

This pirating of her work in the United States made Mrs. Somerville especially sympathetic to the campaign mounted in 1836 by Harriet Martineau (1802–1876) to gain legal protection — and thus royalties — in the United States for the works of British authors sold there. Miss Martineau asked for — and got — Mary Somerville's signature on a petition to the United States Congress 'to obtain a law this Session, to answer to that by w[hic]h we have given to Amer[ns] the right of taking out Patents . . . on the same terms with Englishmen'.[87] Two other letters from Miss Martineau at the time[88] suggest that Mrs. Somerville took an active part in the effort. The petition was carried back to America by a visiting naval officer, Lieutenant Charles Wilkes, with whom Miss Martineau was on good terms, as was Mrs. Somerville. Not for some decades however would reciprocal copyright legislation be passed. In her autobiography Mary Somerville commented drily:

It [*On the Connexion of the Physical Sciences*] went through various editions in the United States of America to the honor, though not to the profit, of the author however the publisher obligingly sent me a copy. . . .[89]

4. New Honours and a New Edition

John Murray in early February commissioned Thomas Phillips, R.A. (1770–1845) to add Mary Somerville's likeness[90] to those of several other distinguished authors — Byron, Scott, Thomas Moore, Southey, Coleridge and Thomas Campbell — in what Murray called his Collection of Eminent Persons. Mrs. Somerville was delighted and immediately wrote Murray to say 'how much I am gratified by the compliment'.[91] She enjoyed the sittings[92] and the quarter-length oil portrait when completed was 'reckoned very like'.[93] When exhibited in May at the Royal Academy one critic called it 'a masterpiece'[94] while another declared that the picture was rendered 'still more interesting as being a resemblance of the most extraordinary woman living, or perhaps who ever lived'.[95] The portrait is now in the Scottish National Portrait Gallery.

This was but the first of a series of attentions and honours bestowed on Mary Somerville in the months immediately following the publication of her second book. Writing her son on 6 March, she remarked, '. . . we are much out. I am a kind of tame Lioness at present, and Martha says that even she & Mary are Cubberized'.[96] Early in April — after some adroit manoeuvering on William Somerville's part and with the assistance of Lord John Russell[97, 98] — Queen Adelaide graciously accepted a copy of On the Connexion of the Physical Sciences.[99] Shortly afterward Mary Somerville had news of her election to honorary membership in the Société de Physique et d'Histoire Naturelle de Genève,[100] and in May of a similar honour paid by the Royal Irish Academy.[101] Jane Marcet[102] and Pierre Prevost were her sponsors in Geneva and Willliam Rowan Hamilton (1805–1865)[103] in Dublin. The Geneva honour had grown out of her presentation of a copy of The Mechanism of the Heavens to the library of the society[104] and the favourable impression the work made on the group. Hamilton, who had met Mrs. Somerville in Cambridge in 1832, was well aware of the high regard in which she was held there and in London and Paris and must have seen her as a happy addition to the roster of the Dublin society.

In mid-summer the long-postponed sittings to Chantrey for the bust for the Royal Society took place. The society had commissioned the work some 30 months earlier but illness, absence, and press of business had delayed matters. Chantrey's likeness was highly praised by viewers and remained in his studio for some years before the marble sculpture finally reached the rooms of the Royal Society. In October 1834, Mary Somerville received an unexpected acknowledgement from the Russian Emperor of the copy of her Mechanism of the Heavens which she had sent him earlier. The 'trinket'[105] he had commanded to be dispatched to her reached Chelsea; it was a magnificent gift of opals set with brilliants.[106]

The Somervilles' old friend, Captain Francis Beaufort (1774–1867; F.R.S., 1814), hydrographer to the navy, was with her when the jewels were delivered.[107] He had come to Chelsea as the bearer of a petition from a wealthy Liverpool merchant and shipowner, William Potter, who craved permission to call his new sailing vessel after the celebrated Mrs. Somerville. Potter wished to give, Beaufort explained, his 'new and fine ship . . . [a] name which shall do equal honor to both the ship and the owner'[108] and to use a copy of Mrs. Somerville's 'bust for her figurehead'.[109] Mary Somerville, with her lifelong and ardent attachment to the sea, was delighted. The launching of the new vessel on the last day of 1834 was reported in the *Times,*[110] which seized the occasion to give at some length remarks made by the Tory M.P. for Liverpool, Lord Dudley Sandon (1798–1882; F.R.S., 1853) and his Tory fellow-candidate, Sir Howard Douglas (1776–1861; F.R.S., 1812). No mention is made in the piece of Mrs. Somerville.

The *Mary Somerville* sailed for Calcutta and Canton on 20 March 1835.[111] She carried a gift of books sent her captain by Charles Babbage. Mrs. Somerville herself sent Mr. Potter a copy of her new second edition. He in return dispatched to her a down tippet, one of a dozen 'received lately from Bengal'.[112] The vessel disappeared on her maiden voyage, lost it was believed in a typhoon in the China Sea.[113]

Despite all the attention, excitement and gaiety that filled these months, neither of the Somervilles abandoned science. William Somerville[114, 115] and Woronzow Greig[116] were among the founding members of the new Statistical Society of London, established in March 1834. Greig became one of the secretaries of the new group[117] and read the first paper ever presented at its sessions.[118] His mother turned again to mathematics. While in Paris, she had begun, at Poisson's suggestion, a piece dealing with 'the analytical attractions of the spheroids, the form and rotation of the earth, the tides of the ocean and the atmosphere and small undulations',[119] about which she had also inquired from Ivory. This new work was intended to round out the *Mechanism of the Heavens,* should there be a second edition. Back in London, she completed this paper and then, as she preferred 'analysis to all other subjects',[120] she undertook the composition of a work 'on curves and surfaces of the second and higher orders'. This 246-page treatise she also finished in 1834. Neither of the two works was ever published.

'While writing this [paper] *con amore*', Mrs. Somerville noted in her autobiography, 'a new edition of the "Physical Sciences" was much needed, so I put on high pressure and worked at both'.[121] By August she was correcting proof-sheets and in December a second edition, dated 1835, was published. The changes made are apparent through comparison of the first two editions. Papers in the Somerville Collection make it possible to trace the origin and course of some of the alterations.

In preparing her new edition Mary Somerville once more had the assistance of several scientific colleagues. In addition she drew upon the comments made by reviewers and in letters of thanks for presentation copies. Babbage[122] sent precise figures for a measurement. Lubbock,[123] as Treasurer of the Royal Society, sent plates depicting acoustical figures from a *Philosophical Transactions* paper by Charles Wheatstone (1802–1875; F.R.S., 1836).[124] W. H. Smyth (1788–1865; F.R.S., 1826), in a series of letters over the six months April through September, conveyed his latest observations on stars and his perceptive comments about astronomical findings and problems.[125] Since Herschel's departure, Smyth had served as Mrs. Somerville's chief astronomical adviser. The Somervilles and the Smyths had much in common and were good friends. Smyth, after an active naval career, had retired to Bedford, where he built himself a fine observatory. His wife, mother of a large young family, was his capable assistant. Smyth himself took an active part in the Royal Society, the Geological Society and especially the Astronomical Society. The observations which he sent to Chelsea in the summer of 1834 were immediately incorporated into the revised text on stars and comets.[126]

Once again Michael Faraday read Mary Somerville's sheets and again some of his comments on them survive among her papers. His note of thanks to her for her present of a copy of the first edition of her work reminds a reader how new were the researches Mrs. Somerville reported:

. . . I cannot resist saying too what pleasure I feel in your approbation of my later Experimental Researches. The approval of one judge is to me more stimulating than the applause of thousands that *cannot* understand the subject.[127]

In March 1834 Faraday completed his eighth series of 'Experimental Researches' — studies of the electricity of the voltaic pile.[128] In April and May he and Whewell carried on their important correspondence about new electrical nomenclature.[129] Mrs. Somerville adopted his ideas about electricity but not his new vocabulary; she continued to use terms such as 'negative pole' and 'positive pole'. The single surviving sheet of Faraday's comments on the second edition is dated 'August 1834'[130] and is directed to her Section XIV,[131] dealing with forces and matter. Faraday addresses five specific topics: definite proportions, atomic weights, definite proportions of electricity (Faraday's laws of electrolysis), crystallization, and the density of the atmosphere. The printed text of her second edition (pages 129–142) reveals that all his advice was taken. Again, no manuscript for this edition exists in the papers.

Another of Mrs. Somerville's consultants during the preparation of this edition was Arago, in Britain to attend the fourth meeting of the British Association. Mary Somerville and her husband were urged by Scottish and English friends to join the philosophers who flocked to Edinburgh in September 1834. One letter of invitation is especially noteworthy, as it expresses the real but little-voiced concern many Scottish intellectuals felt as the days of the Scottish Enlightenment passed. Its writer is Francis Jeffrey (1773–1850) —since July Lord Jeffrey — who pressed Mrs. Somerville to come to the meeting:

. . . We think we have a right, both to you and Dr. S., as Scotch — & beg you to consider whether you ought not to make some sacrifice rather than deprive our first national celebration [in science] of the honor and benefit of your attendance — I must not speak of our *Poverty* to strangers. But you & I know it too well to have any scruples on the subject — & now that we have no Stewarts — or Robisons — or Playfairs to make a shew with — we really cannot afford to be without a name such as yours. . . .[132]

In fact no Edinburgh professor read a paper at the meeting, a decline in Scotch science remarked by the *Athenaeum*.[133]

Mrs. Somerville was busy with her second edition and remained in London.

Her situation in 1834 is in sharp contrast to that of the hundreds of scientific men — mostly provincials — who flocked to these annual meetings. For two decades she had been granted easy access to the foremost scientific minds in the nation. She could easily acquire any scientific information and introduction she needed from leading practitioners at home and abroad. She had the constant stimulus of good scientific society. Since she neither had —nor reasonably could expect — any official position in science — any office or professorship or fellowship — she was immune from the feelings of neglect or frustration that beset many B.A.A.S. members. Her needs were greatly different from theirs and her goal — the advancement of science — identical. She would seek to do so through her exposition of scientific work.

Among the foreign philosophers in Edinburgh the energetic and charming Arago attracted the most attention. After the meeting was over he stayed on in Britain for some weeks. While in London he saw the Somervilles frequently[134, 135] and knew that she was busy with the second edition of her work. He brought to her attention some recent discoveries about radiant heat made by his Italian protégé, Macedonio Melloni (1798–1854; F.R.S., 1839), who had been a political exile in Paris for some years. Radiant heat was a subject of great interest at the time and Melloni's researches were highly regarded. It was too late to include the information Arago brought in her text but Mrs. Somerville did add a six-and-a half page 'Supplement' in order to give a full account of this 'important and interesting' new work.[136]

Her revisions for the new edition — mostly addition of material, correction of errors, and polishing of prose — were extensive. The basic plan of the book remained unchanged; its execution was vastly improved. A table of contents showing the material covered in each 'Section' — of which there were still thirty-seven — and a much fuller index made the volume more useful to its readers. Corrections too late for the first edition were now made in the second. Its text was carefully scrutinized and even minor points put right. 'Mr. Faraday', for example, was now 'Dr. Faraday'; in 1832–33 she had overlooked the degree given him at the second B.A.A.S. meeting. The portion called 'Explanation of Terms' in the first edition became a much fuller segment labelled 'Notes' in the second. Some of its better-known definitions were dropped, many obscure ones added.

The most striking change was the inclusion of a great many figures and four pages of plates. Some 73 diagrams, drawings, and geometrical figures appeared in the 'Notes' to assist in explanation. There were sketches of apparatus in some instances. Four pages of plates depicting 60 different acoustic figures were added. Brewster's criticism of 'want of ocular delineation' was taken seriously; the acoustic figures were the ones he had recommended in his review.[137] The book was still 'popular' in that it contained no mathematical presentations. Mrs. Somerville at Murray's urging had sought — not very successfully — to use plainer language and simpler explanations.

He advertised the work as 'A new edition, most carefully revised, particularly with the view of simplifying the work, and rendering it intelligible to unscientific readers, with numerous illustrative woodcuts'.[138] Murray printed

3000 copies, a thousand more than for any other edition. Of these, eleven went to Stationers Hall, three to reviews, and forty-three to the author.[139] Letters in the Somerville Collection account for thirteen of these. Once again a copy was presented to the Royal Society.[140] Another was sent to the London Mechanics Institution.[141] The price was increased to 10s.6d. a copy.

Cost of the edition was £466.17.2, of which £33.14.8 went for advertising. Murray placed six advertisements of the new work in the *Literary Gazette*[142] between 20 December and 11 April, three in the *Athenaeum*[143] from 27 December through 21 March. Although dated 1835, the book was available in December 1834. Sales were so good that Murray's statement to Dr. Somerville the following June showed '2954 [copies out of 3000] sold for £1008.16.6'. The author's share of the profits at that date was £361.6.3, Murray's, £180.13.1.[144] Once again Mary Somerville — and John Murray — had scored a scientific and publishing triumph.

5. Mary Somerville and a Few Scientific Women

As a girl Mary Somerville had encountered strong opposition to her pursuit of mathematics but she had persevered. With William Somerville and among Scottish philosophers she found encouragement and help for studies unusual for women at the time. In London she was fortunate enough to move immediately into a circle that accepted and approved of women's scientific interests. At a time when science was not yet professionalized, when no economic threat was posed to males by female cultivation of science, eminent scientific practitioners, eager for converts to the cause of science, welcomed both men and women. Throughout her London years Mrs. Somerville's scientific circles contained several scientific ladies other than herself.

Her relations with the wives, children and families of her British scientific colleagues were uniformly friendly and in most instances warmly affectionate. There were constant visits, letters, interest in daily happenings and in the health and well-being of each other and of their families. With the five wives of her first scientific circle — Jane Marcet, Mary Frances Kater, Eliza Young, Anne Chantrey and Mary Blake — Mary Somerville shared an interest in society, in domestic matters, in children and in entertainment. Two of the five — Mrs. Marcet and Mrs. Kater — actually did scientific work of a kind, Jane Marcet[145] with her 'Conversations', Mary Kater assisting her husband with his calculations and results, but neither thought of herself as scientific.

The Somervilles' second inner scientific circle was larger, its ladies more numerous and more active. Three of the group — Charlotte Murchison (1789–1869), Mary Morland Buckland (1797–1857) and Mary Elizabeth Lyell (1808–1873) — were practicing geologists, accompanying their husbands on field trips, making observations, sketches and models, and assisting in the collection of specimens and the preparation of papers. Among the Somervilles' larger circle Annarella Smyth (married 1815) was an enthusiastic assistant in the astronomical observations of her husband W. H. Smyth while she reared a young family, several of whom became scientists or married scientists.[146] Henrietta Beaufort (1778–1865), sister of the hydrographer Francis Beaufort, was the anonymous author of two little scientific works, *Dialogues on Botany* (1819) and *Bertha's Journal* (1829), and often served as her brother's scientific secretary.[147] These ladies were in no sense rivals for scientific eminence but rather enjoyed scientific pleasures together and scientific conversation. Charlotte Murchison was the leader in opening Lyell's geological lectures at King's College to women[148] and Mrs. Somerville and her daughters had soon joined her there.[149] Mary Somerville's pre-eminence was acknowledged by all her female contemporaries; they looked upon her in most cases as uniquely gifted.

The Somerville papers — with one slight exception[150] — offer no evidence that Mrs. Somerville associated with three other scientific ladies of the time — Mrs. Sarah Bowdich Lee (1791–1856), Mrs. Jane Loudon (1807–1858) and Mrs. Maria Gray (1787–1876) — all gifted naturalists. One passing reference to Mrs. Lee is the only acknowledgement of any of them.

Of these ladies, only the last three and Mary Somerville were widely known as 'scientific'. None in the large group, except Mrs. Smyth, was at all mathematical. Mrs. Somerville's interests were rarely found among scientific women, most of whom at this time tended to favour botany, natural history or geology.

With the success in such close order of her first two books, however, Mrs. Somerville became in the public eye *the* authority on all matters pertaining to women and science. Not only was her opinion solicited as to whether ladies should attend the reading of papers at the British Association, but she was on occasion consulted by parents and guardians facing the vexing problem of whether a young girl could with safety be allowed to follow mathematical and scientific inclinations. Mrs. Somerville of course had done so with splendid results. At the same time she was the careful mother of daughters and therefore apt to be sympathetic to parental anxieties. In some cases, having assented to mathematical studies, the inquirer asked for guidance in such studies or for the recommendation of a tutor.

Soon after the publication of *On the Connexion of the Physical Sciences*, Hudson Gurney (1775–1864; F.R.S., 1818) — a liberal and well-educated man with scientific interests — wrote Mary Somerville a letter typical of this sort of appeal. He explained that his wife's niece

. . . instead of adhering to the straight course . . . ordinarily taught by the instructors of *Miss*-ery . . . [wished also] to read *mathematics* . . . I have always been for her attempting it — if she had the inclination . . . as it is always . . . good to know something *about* things — The Lady who has been instructing her has been very much *against* this fancy. But in *Norfolk* she has been reading Euclid with a Mr. Saint . . . [who] gives her the Character of being one of the aptest Scholars he ever had . . . I take the Liberty of asking you if you know of any person, who gives Lessons, with whom she could continue her reading when in town . . . I thought you would forgive me . . . [for taking liberty] in asking for Instruction from the *Fountain Head*.[151]

Some years earlier a similar query — and one more pregnant with consequences — had come to Mary Somerville from Lady Noel Byron (1792–1860), the widow of the poet. She and Byron had parted in January 1816 when their only child, Augusta Ada (1815–1852) was but a month old. Since that time great public curiosity had surrounded both Lady Byron and her daughter, who was carefully and attentively reared in an atmosphere excessively protective because of their peculiar circumstances. Ada Byron was high-strung and delicate, her nervousness being attributed to 'a complaint connected with the spine'.[152] Because she was so frequently ill, indeed at times an invalid, her education was sporadic but always carried out under the direction of her mother and according to whatever 'system' Lady Byron currently favoured. Emphasis, however, was always on character formation.

As a girl Lady Byron had read mathematics with William Frend (1757–1841), the Cambridge don who had been expelled from his fellowship in 1793 for reformist views, and kept up an admiring friendship with her old tutor and his family. She early discerned a mathematical bent in her own daughter and did not discourage it. She was convinced that mathematics would remedy 'a want of order'[153] she observed in the child.

Some time around 1830 — the Byron ladies rarely date letters fully —Lady Byron took a house, Fordhook, near Acton in Middlesex, which became one of her favourite residences. She and the Somervilles, in neighboring Chelsea, were soon acquainted; they had a number of good friends in common, including Joanna Baillie and Maria Edgeworth. Woronzow Greig later wrote, in a fragmentary memoir found in the Somerville Collection, that his

. . . first recollection of Ada Byron was about 1832 or 3 . . . when as a young girl she was a visitor at the house of my mother at the Royal College Chelsea . . . She was very intimate with my half-sisters who were about her own age and as she had even in these early years a decided taste for science which was much approved of by Lady Noel Byron she took every opportunity of cultivating my mother's acquaintance. . . .[154]

By that date young Greig was living in London chambers and not always current with day-to-day happenings at Chelsea.

The earliest of the dozens of letters from the Byron ladies in the Somerville Collection appears to be one from Lady Byron welcoming the 'friendly but judicious suggestions' which Mrs. Somerville had made 'respecting . . . [Ada's] objects and dispositions'. Lady Byron wrote that she found a 'remarkable agreement of some' of Mrs. Somerville's opinions and her own and was gratified to enlarge Ada's

. . . acquaintance . . . under the protection of one who would assist her in keeping her balance of mind. The pursuits of science have been very beneficial to her in that respect. You have the happy effect of increasing her interest in them, whilst you are able to moderate any excessive or exclusive attention to that line of study.[155]

Ada herself found Mrs. Somerville and her daughters delightful and they liked her. An affectionate friendship sprang up between the two households that continued to death. Mary Somerville lent Ada books, advised her on her studies, set mathematical problems for her and helped with their difficulties, and above all talked mathematics to her. When the Somervilles went to Paris in the autumn of 1832, Ada was on the verge of entering London society; her first London season was 1832–33, culminating in her presentation at Court on 10 May 1833. On the Somervilles' return to Chelsea, the intimacy was immediately resumed. Mrs. Somerville and Ada still talked mathematics but the young ladies now had the topic of balls and gowns and fashionable society to add to their conversation.

Since her separation from the poet in 1816 Lady Noel Byron had not gone formally into society. Now that Ada was 'out' the problem of finding a chaperon — one acceptable to her mother, who through the years had learned to be wary of associations involving her daughter, and congenial to the young lady — was a constant one. In Mary Somerville they found the ideal solution,[156] a lady completely liked and trusted by both and part of London intellectual society.

During 1834 and 1835 visits and letters between Fordhook and Chelsea were frequently exchanged. There were expeditions to see the Hydro-Oxygen Microscope in Bond Street, to attend lectures at the Royal Institution, and often trips to Charles Babbage's for talk or to his evening parties. One letter in the Collection suggests that Miss Byron may have met Babbage before the Somervilles went abroad,[157] but it is certain that they were on friendly terms soon after the Somervilles returned. Woronzow Greig in his fragmentary memoir written after Ada's death notes that

. . . she constantly accompanied my mother and sisters to Babbage's evening parties which were then the fashion in London, and thus she commenced an acquaintance with a man who for his scientific pursuits and reputation acquired an influence over her which eventually did her much harm.[158]

Since Greig was a good friend of all parties — Babbage, Ada Byron and her husband, William King, later Earl of Lovelace — and since Greig played a central role in the last unhappy days of their relationships, he speaks with authority. Those days, however, were far in the future when the three young ladies and Mrs. Somerville were so much in London society in 1834–35.

Ada Byron continued her mathematical studies with Mrs. Somerville during these months. By mid-November 1834 she reported herself as 'going on very well indeed . . . [having reached] Monsieur Legendre'.[159] Mary Somerville found in her a pupil worthy of the teacher. Her marriage in July to Lord William King (1805–1893; F.R.S., 1841), who was created Lord Lovelace in 1838, strengthened the bonds between the families. King and Woronzow

Greig had become friends at Cambridge and continued on a basis of close intimacy. The two made the grand tour together in 1827–29 and had met Babbage in Rome at that time. In London they saw each other frequently. Greig in fact took credit for the match between his friend and Miss Byron.

After the marriage, the Somervilles were several times at Ockham Park, King's estate in Surrey. On occasion Babbage and Woronzow Greig were also guests. Ada worked sporadically at mathematics but also turned her attention to other hobbies and interests almost capriciously. After a pleasant Christmas visit there, Mary Somerville wrote her daughter that she thought Ada's taste for mathematics was wearing out.[160] Nevertheless Lady Lovelace throughout her short life kept in touch with Mary Somerville.

After the Somervilles left for Italy, Ada Lovelace turned to Babbage and Augustus De Morgan for mathematical instruction. De Morgan on one occasion made a confidential comparison of his pupil's mathematical powers to those of Mrs. Somerville.[161] He concluded that Ada showed more mathematical genius. Mary Somerville would rejoice in the genius and deplore the lack of steady application, but her affection for her friend would remain undiminished.

In the coming decades there would be steady pressure for women's education and Mary Somerville's pioneering accomplishments would be widely hailed. In her own time she did whatever she could to promote learning among women and open opportunities to them.

CHAPTER 8

THE CIVIL LIST AND MARY SOMERVILLE

In 1835 Mary Somerville received the most solid recognition of her career. It came not from her friends the Whigs but from a Tory ministry. Sir Robert Peel's reasons for placing her on the Civil List in April 1835 reflect not only her public position at the time but changes that were taking place in the award of pensions and in politicians' perceptions of the place of science and scientists in the national life. Further, a close examination of this particular event not only discloses some of the political implications which were read into it at the time and some of its after-effects but the hairbreadth margin by which her award was officially validated. Of interest also is a comparison of factors affecting her pension and its size with those governing some others given for scientific and literary work.

Ever since the Whigs had come to power in November 1830, the Somervilles had hoped for some minor favour from their friends in the government, especially from Lord Brougham, who could easily do something for the young barrister Woronzow Greig.[1] As the months passed, however, with no more than hints and half-promises from the Lord Chancellor, they gradually and resentfully came to expect nothing from that quarter.[2] Neither the publication of *The Mechanism of the Heavens* nor of the *Connexion of the Physical Sciences* — both works in which Brougham had displayed marked private interest and which were widely hailed — brought any public notice from him, though he urged the claims for governmental largesse of many other liberal philosophers. Indeed, when the end of the Whig ministry came in sight late in 1834 — after a series of embarrassments over vexing questions of civil and religious tolerance — he could declare that only two of the clergy 'who deserved well of the liberal party' were still unrewarded; he lamented the fact that 'he had been unable to provide' anything for Adam Sedgwick and the Cambridge historian Connop Thirlwall (1797–1875).[3] Even that failure was remedied, for at the last moment a prebendary stall at Norwich and a living in Yorkshire fell vacant through death, and Brougham was able to present the first to Sedgwick and the second to Thirlwall on the day before surrendering the Great Seal. It is clear that Brougham considered any claims of the Somervilles — to whom in fact he did have a moral obligation — to be among his least pressing. Any speculation about what he might have done in future is meaningless, since he never returned to office.

After the King dismissed Lord Melbourne (1779–1848; F.R.S., 1841) in December 1834, he asked Sir Robert Peel to form a government. Peel did so, but opposition to a Tory ministry was so strong that it was decided to go to the country. Parliament was dissolved on 30 December, and in the elections of January 1835 the Tories won a majority but not enough seats to defeat any coalition of Whigs and O'Connell's Irish party. Peel continued in office but in daily peril of defeat.

With the advent of the new year, monies for additional civil pensions were once more at the disposal of the government, and Peel was eager to take advantage of this patronage for the good of the country and the good of the Tory party before he was forced from office. He had long seen the civil list as more than a mere device for recognizing military, diplomatic or political services; he believed that it could and should be used as an instrument for awakening and supporting broader national cultural concerns. In the 1830 debate in the Commons over the list — which came about, as was customary, through the accession of a new monarch — he had argued strongly for its reform and for restricting awards to those who 'either have rendered a service to the public or who are nearly related to those who have, and to enable those who have rank and title but no means of maintaining them to live at least in public dignity'.[4] This debate resulted in the reduction of the civil list, correction of some of its more flagrant abuses, and a 'resolution'[5] which limited the total paid out annually to £75,000 and restricted the amount given in new awards in any one year to £1,200. It was this last sum that Peel in 1835 was eager to distribute well and soon.

He had long been interested in science and its advancement and was a discriminating judge of scientific merit. He regarded with favour the steps the Whigs had taken in moving beyond military, diplomatic or political service to recognition also of science and scientists. On assuming office late in 1830, the Whigs had embraced Babbage's notion — presented some months earlier in his *Reflections on the Decline of Science* — that pensions and other governmental awards for philosophers would strengthen English science.[6] Their actions in this regard had met, as Peel knew, with general approval. During the Whigs' first year in power, seven knighthoods had gone to scientists. John Rennie was given a baronetcy after the completion of the new London Bridge — a gesture of gratitude also to his late distinguished engineer father John Rennie — and in the autumn of 1831 six 'scientific knights' — Herschel, Leslie, Charles Bell, Harris Nicolas, Brewster and Ivory[7] — were added to the Guelphic Order. Ivory's civil pension — £100 which he had been given in 1819 — was raised to £300 annually and Brewster received £100 — actions initiated, as most of these were, by Brougham. A grant of a like amount was made, at the behest of a group of geologists, to the venerable William Smith (1769–1839; F.R.S., 1806) and announced at the Anniversary Dinner of the Geological Society in February 1832.[8] In Cambridge the following year at the B.A.A.S. meeting, one of £150 annually was presented to John Dalton (1766–1844; F.R.S., 1822).[9] Never had science been so frequently recognized by government, and the Whigs took full credit. Almost unnoticed was the fact that most the awards had gone to strong Whig supporters among scientists.

Peel was determined that the Tories, at this first opportunity to act on pensions since the civil list reforms of 1830, should do no less for science and literature. When he, as Home Secretary, had written Sir James South on 10 July 1830 of the King's intention to give South a baronetcy and an annual pension of £300, he had not only cited South's 'signal service to practical navigation' as a reason but had stated that he himself was 'anxious . . . to relieve the country from a charge of perfect indifference to subjects of scientific nature'.[10] Now as Prime Minister he had opportunity to continue this endeavour. On 31 January 1835, shortly after resuming office, he received a letter from the King assenting to his views about 'the propriety and expediency of rewarding and giving encouragement to men distinguished by their literary talents and scientific attainments' and 'highly' approving of his recommendations to make the long-time secretary of the admiralty, John Barrow (1764–1849; F.R.S., 1805), and the poet Robert Southey baronets.[11] Although Barrow had been active in the management of the Royal Society since early in the century and had helped organize the Royal Geographical Society in 1830, his scientific work was but a minor factor in the elevation Peel proposed; Barrow had a long record of public service, was in the special good graces of the King (whom he had served at the admiralty), and was a strong Tory supporter. Since 1806, he had enjoyed an annual pension of £1,000. Peel's coupling of his proposed baronetcy with 'scientific attainment' was merely a useful hint of future intentions; even without contributions to science Sir John would probably have received a title in 1835. Peel's personal inclinations and political shrewdness went further and persuaded him that merit rather than mere partisanship might well be the first element in choosing scientific recipients.

Peel consulted his friend and supporter, John Wilson Croker (1780–1857; F.R.S., 1810), about matters of new pensions. Mary Somerville's name may have come up during this or subsequent conversations between the two. A letter from Croker to Peel, undated but obviously written after mid-January 1835, suggests that hers may have been one of the first names mentioned:

My dear Peel,

Let me remind you of the pension to Mrs. Somerville. I never saw her, and have no kind of interest in the matter, but as it concerns the honour of your Administration, and the cause of science and letters. I have made such enquiries about her as I could venture to do without exciting suspicion as to my object or leading by-and-bye to a suspicion that I was the benefactor, who, in fact, only ring the bell. She is the daughter of an Admiral Fairfax, who was Lord Duncan's Captain in his victory at Camperdown, brought home his dispatches, and was knighted. He is dead about twenty years. She married first the son of Admiral Greig, of the Russian service, by whom she had a son, who has all his father left. She married secondly Dr. Somerville, who is physician to Chelsea Hospital, with no means, I am told, but his salary. She has two daughters, who, with herself, are unprovided for, except by the doctor's situation.

I ought to tell you that I heard a whisper that Brougham had promised to do something for them, and that they think he played false with them, but I know nothing of the details and did not choose to enquire. 200 l. or 150 l. a year would surely be well applied in this case; or say 150 l. to her and 25 l. each to the daughters. The child and grandchildren of Sir W. Fairfax have a degree of merit exclusive of Mrs. Somerville's literary reputation. . . .[12]

Peel's reply to Croker was a single long sentence: 'As far as the *abstract* case is concerned, I have both ample means and equal inclination to give a pension to Mrs. Somerville, but there are three or four matters connected with this, and with aid to literature, that I should like to speak to you upon before I do anything'.[13] A week later he wrote that in the whole matter he felt that he 'must be very cautious not to confine pensions to Whigs or Liberal professors of literature . . .'.[14]

Croker was a frequent contributor to the Tory *Quarterly Review,* and, after the election, called on J. G. Lockhart, its editor, and John Murray, its publisher, to say that 'Peel was anxious to know if . . . [they] had anything to suggest to the new Minister for the department of Literature and Science'. Lockhart, writing of this occasion to his and the Somervilles' friend H. H. Milman (1791–1868) in 1843, reported that 'Murray said he wished there could be a pension for Mrs. Somerville . . .'. Lockhart, with Murray concurring, suggested some 'London preferment' for Milman (he was made canon of Westminster and rector of St. Margaret's) and a living for the poet George Crabbe's (1754–1832) son George (1785–1857) (who received two vicarages in Suffolk).[15]

Peel's first approach, however, was not to Mary Somerville but to G. B. Airy, Plumian professor of astronomy and director of the Cambridge Observatory. The previous October Airy, who was considered sympathetic to the Whigs, had provisionally accepted a Whig offer of the post of Astronomer Royal but had delayed, because of immediate Cambridge responsibilities, entering formally into office. The change of ministry upset this arrangement and the place remained vacant. On 19 January 1835, Airy, who had long been associated with Trinity College, Cambridge, and had a good sense of the 'Trinity network', unsuccessfully petitioned the new Lord Chancellor, Lord Lyndhurst (1772–1863; F.R.S., 1826) for a Suffolk living for his brother William Airy. Lyndhurst, who in 1790 had been second wrangler and was later a fellow of Trinity, had been for well over a decade high in the councils of the Tory party. In his autobiography Airy wondered whether it was Lyndhurst who brought his name to Peel's attention at this time or whether Peel, whom he had met the previous year at a dinner given by the Duke of Sussex, recollected him from their conversation on that occasion.[16] The Cambridge astronomer was much in the public eye of the scientific community early in 1835, for he had been given the Lalande prize by the French Institute the previous November and on 9 January was elected a correspondent of the French Academy.

On 17 February Sir Robert Peel, without previous warning of his intention, wrote to Airy:

In acting upon the Principle of the Resolution [on civil pensions] in so far as the Claims of Science are concerned, my *first* address is made to you, and made directly, and without previous communication with any other person, because it is dictated exclusively by public considerations . . . I consider you to have the first claim on Royal Favour which Eminence in those high Pursuits to which you life is devoted, can give . . . I make it [the offer of a pension of £300 annually to you or to your wife] upon public grounds, and I ask you, by the acceptance of it, to permit the King to give some slight encouragement to Science, by proving to those who may be disposed to follow your bright example, that devotion to the highest Branches of Mathematical and Astronomical Knowledge shall not necessarily involve them in constant solicitude as to the future condition of those, for whom the application of the same Talents to more lucrative Pursuits would have ensured ample Provision.[17]

On the following day, 18 February, Airy wrote Peel, accepting the offer and asking that the pension be given Mrs. Airy. Sir Robert on 19 February directed that a Warrant be prepared granting her £300 annually from the civil list for her lifetime.[18] The Prime Minister, customarily prompt in executing tasks, lost not a moment in this matter; he was well aware that he might have to leave office at any time. Airy, on the other hand, in his decision to give the pension to his wife, demonstrated a breathtaking disregard of common sense. A cool and methodical man, well acquainted with the laws of probability, he made in this instance what appears to be a wildly romantic gesture. At the time he was 34 years old, his wife 31. They had married in 1830 and, in a period when death from childbirth was frequent, already had three children, with six more yet to come. One of Mrs. Airy's sisters had died of consumption since their marriage, another was gravely ill of the disease at the time of the Peel correspondence (and died in April 1835). Nevertheless, Airy directed that the pension go to his wife 'for her lifetime'; she did, in fact live until 1875, receiving a total of £120,000, but Airy survived her by six years.

The Whigs, hearing of Peel's grant to Mrs. Airy, were not disturbed. Rather they took it, when the 'subject came before Parliament . . . [as] vindicating their own propriety in having offered . . . [Airy] the office of Astronomer Royal in the preceding year . . .'[19] and, after they returned to office in April, renewed their proposal. Airy accepted and in October 1835 became Astronomer Royal. In his case, Peel seems to have been motivated by his own good opinion of Airy and his work and by the high regard in which the astronomer was held by competent colleagues, including French colleagues. It is to Peel's credit that he valued the cool judgement of a man's peers.

The matter of Airy's pension was settled some six weeks before Peel approached Mary Somerville. Early in this interval, however, the public learned of an unprecedented honour given her, one that could but strengthen any claim on Royal favour. At the Anniversary Meeting of the Royal Astronomical Society on 13 February, she and the venerable Caroline Herschel were unanimously elected 'Honorary Members of the Society'.[20] Never before had any woman been named to fellowship in this specialist society and some years would pass before another was chosen. Not only were the two new members gratified and honoured by the distinction they received but they were pleased to be coupled. Caroline Herschel, sending her thanks from Hanover, remarked that she only regretted 'that at the feeble age of 85 I have no hope of making myself deserving of the great honour of seeing my name joined with that of the much distinguished Mrs. Somerville'.[21] Mrs. Somerville wrote Miss Herschel that in her own mind the Astronomical Society had greatly added 'to the value of that distinction [her election] by associating my name with yours, to which I have looked with so much admiration'.[22] The event was reported in the daily papers[23] and in the *Literary Gazette*[24] and could not have escaped Peel's notice.

Likely he was aware also that on 2 March she had been received, in private audience, by the heiress to the throne, the Princess Victoria, and her mother, the Duchess of Kent. On 25 February the Duchess's chief attendant, Sir John Conroy, had written Dr. Somerville that they would have 'the greatest pleasure' in seeing her at Kensington Palace.[25] On that occasion the Royal ladies were 'generous and amiable',[26] received her 'very graciously, and conversed for half an hour'.[27] Mrs. Somerville, after asking Ada Byron to make enquiries as to the proper inscription and receiving her reply,[28] presented a copy of the second edition of the *Connexion of the Sciences* to Princess Victoria.[29]

On 30 March the Prime Minister wrote Mary Somerville, in his own hand, a gracious and flattering letter which echoed many of the sentiments and words of his earlier missive to Airy. In it he offered her, without obligation on her part, an annual pension of £200. She had not anticipated this 'joyful and unexpected event',[30] for Peel in his letter declared that he preferred 'making a direct communication to you to any private inquiries into your pecuniary circumstances as to my proposal through a third party'. He asserted, as he had to Airy, that 'among the first claims on the Royal favour [are] those which are derived from eminence in science and literature' and that his object was 'a public one, to encourage others to follow the bright example which you have set, and to prove that great scientific attainments are recognised among public claims. . . '.[31] Mrs. Somerville immediately and gratefully accepted, calling his conduct 'liberal and . . . disinterested'.[32] On 6 April a King's Warrant was issued, giving her a pension of £200 annually beginning on 1 April 1835.[33] Two days later, on 8 April, the government was defeated over a question of Irish reform and Peel resigned. Mrs. Somerville's pension, however, was safely hers.

Sir Robert had written her on 1 April to say that he had 'this day had the satisfaction of completing the Instruments . . . [to execute] the intention communicated to you in my former letter' and to thank her for the copy of her second edition of the *Connexion of the Sciences* she had just sent him. He assured her that it would 'be highly prized by me, though it will not have the charm of novelty, for with the contents of it, at least of the first edition of it, I am quite familiar'.[34]

In January Croker had suggested £200 annually for her, a large pension for a woman and a generous one for a scientific person. Less than Airy's, it was still more than Dalton, Brewster and several others were receiving at the time. Literary ladies were generally given £50 or £100 — Miss Mitford (1787–1855) was given £100 a year in 1837 — and there was no precedent for scientific ladies. Peel may have consulted the Somervilles' old friend Henry Warburton about the amount, for in the following July Mrs. Somerville referred in two letters to his part in the pension matter. In the first instance she described Warburton as having 'ever been friendly, and on no occasion more than when Sir R. Peel consulted him about my pension'.[35] In the second she told of thanking Warburton on the previous day 'for the kind part he had taken in the House when consulted on the subject by Sir R. Peel'.[36]

Not all of Peel's nominees were Whigs. On 4 April he wrote the poet Robert Southey (1774–1843), who had long since exchanged his republican ideas for Tory ones, that he had already executed a Warrant adding £300 annually to the £160 Southey had received since 1806. Peel's four-page letter, again in his own hand and less formal than his communications to Airy and Mrs. Somerville, makes no mention of a baronetcy but does describe some of the factors guiding his disposal of civil list patronage. Deploring 'the miserable pittance' available for 'the Reward and Encouragement of Literary Exertions' — a point he had first raised in the 1830 debate on pensions — Peel declared that he made his nominations

. . . on public grounds, and much more with the view of establishing a *Principle* than in the hope, with such limited means, of being enabled to confer any benefit upon those I shall name to the Crown worthy of the Crown, or commensurate with their claims . . . You will see in the position of public affairs a sufficient Reason for my having done this [executed the Warrant for Southey's pension] without delay and without previous communication with you . . . I have granted the pensions on a principle — on a public Principle, namely the Recognition of Literary and Scientific Eminence, as a public Claim. The other persons, to whom I have addressed myself on this subject are Professor Airey [sic] of Cambridge, the first of living Mathematicians and Astronomers — the first of this country at least — Mrs. Somerville Sharon Turner and James Montgomery of Sheffield. . . .[37]

The historian of Icelandic literature, Sharon Turner (1768–1847), was given £200 annually, while the Sheffield newspaper editor and hymn-writer, James Montgomery (1771–1854), whose lectures on poetry at the Royal Institution in 1830 and 1831 had attracted favourable notice, was awarded £150. A letter from Peel to Croker also mentions a pension to 'a Mrs. Temple, whose husband an African traveller died at Sierra Leone'.[38]

Peel in his little more than three months in office, had disposed of the full amount available for civil pensions in 1835 and had done it in a fairly evenhanded way: £500 to science (Airy and Mrs. Somerville), £650 to literature (Southey, Turner and Montgomery) and £50 to the widow of an African traveller. His political balance was also notable. Airy and Mrs. Somerville were known to be sympathetic to the Whigs, Southey and Turner to the Tories. Peel must have taken special satisfaction in rewarding Mary Somerville, since Brougham — forever flaunting his scientific expertness — had so arrantly neglected her. Indeed the size of her pension may well be related to her association with the former Lord Chancellor. Peel was a large-hearted and large-minded man, but he was well aware, as his letter to Croker shows, of the political value of the power of patronage. Mary Somerville had been highly recommended to his consideration. He could easily have persuaded himself that it would be unfair to deprive her of merited recognition simply because a pension from a Tory prime minister could be interpreted as a slap at Brougham and the Whigs. Its size, while it might emphasize this aspect of his choice, also asserted his devotion to science and his readiness to recognize scientific merit in either sex.

Melbourne, on the other hand — worldly, experienced and tolerant — seems to have seen the awards to Airy and Mrs. Somerville as a political stroke on the part of Peel. He and the Somervilles had been acquainted for almost a decade and a half, but his interests were literary and political rather than scientific. It is clear from several statements he made in October and November 1835 that he was well aware of Peel's action in the two cases and he seems to have taken it as a subtle — if impudent — comment on the neglect of worthy scientists by those who — as a party — professed most loudly their championship of science and scientists.

The connection between Peel's award of pensions to Airy and Mrs. Somerville and Melbourne's handling of the Faraday pension has hitherto gone largely unnoticed. Peel, trained in mathematics at Oxford, felt affinity for work done by Airy and Mary Somerville but he was aware also — he received the *Philosophical Transactions of the Royal Society* and had read Mrs. Somerville's new work on physical sciences — of the growing importance of electrical researches. On 20 April 1835, Sir James South wrote Faraday that it had been Peel's intention, had he remained in office, to give Faraday a pension as soon as possible. The first reply the tender-conscienced Faraday drafted to South, on 23 April, stated that while he approved of such rewards to science, he himself would prefer 'to live by my labours'.[39] His father-in-law persuaded him to modify this stand, but no more was heard of the matter until late October 1835. On 25 October Faraday had what became a notorious interview with the prime minister, Lord Melbourne. During their talk, Melbourne reportedly said that he himself 'looked upon the whole system of giving pensions to literary and scientific persons as a piece of humbug'.[40] Faraday excused himself. That evening he left a note with his card at the Prime Minister's office:

> The conversation with which your Lordship honoured me this afternoon, including as it did, your Lordship's general opinion of the general character of the pensions given of late to scientific persons [i.e., Airy and Mary Somerville], induces me respectfully to decline the favour which I believe your Lordship intends for me; for I feel that I could not, with satisfaction to myself, accept at your Lordship's hands that which, though it has the form of approbation, is of the character which your Lordship so pithily applied to it.[41]

The story became known, reached the King's ears, and William IV, much moved by an account of Faraday's early struggles, declared that the scientist deserved the pension that Peel had intended and that if the Whig cabinet would not give it, then he himself would do so. Public comparison between the actual patronage of science by Whigs and Tories began to appear in the press. The December number of *Frazier's Magazine* asserted that the Whigs had done 'nothing for science or for letters; and there is no reason to suppose that they ever meant to do anything'. Sir Robert Peel, on the other hand, 'had sought out . . . persons whose sole claim upon his notice was the eminence to which they had attained in . . . science or . . . literature'. His disregard of their political allegiance was hailed, Airy and Dalton being named as 'bitter Whigs'. Of Mary Somerville, the writer remarked, 'We say nothing of Mrs. Somerville for a lady's politics go for little; and the lady in question well deserves to find favour in the eyes of men of all parties'.[42] Part of this account — which went into great detail about Faraday's meeting with Melbourne — was picked up by *The Times*, which printed an extract on 28 November 1835.

Faraday, already incensed by Melbourne's words, was grieved at the difficulties that the matter was making for his friends — and these included Mrs. Somerville, whose name had been drawn publicly into the discussion of the political overtones of scientific pensions — but he was unwilling to withdraw his letter. Only a written apology from the Prime Minister would suffice to remove what Faraday saw as a slur on science and scientists.[43] Melbourne responded handsomely, declaring in a letter dated 24 November that he should have abstained from his implied censure of the Tories 'upon such an occasion' and that his

. . . observations were intended only to guard myself against the imputation of having any political advantage in view, and not in any respect to apply to the conduct of those who had or hereafter might avail themselves of a similar offer. I intended to convey that, although I did not entirely approve of the motives which appeared to me to have dictated some recent grants, yet that your scientific character was so eminent and unquestionable as entirely to do away with any objection which I might otherwise have felt, and to render it impossible that a distinction so bestowed could be ascribed to any other motive than a desire to reward acknowledged desert and to advance the interest of philosophy . . . [I hope] that I shall have the satisfaction of receiving your consent to my advising His Majesty to grant you a pension equal in amount to that which has been conferred upon Professor Airy and other persons of distinction in science and literature.[44]

A close reading of this passage reveals that Melbourne had not abandoned his views that the Tories had made political capital out of the pensions they awarded Airy and Mrs. Somerville. His failure to mention her by name may have arisen from delicacy.

For the next two months the Faraday affair was a matter of public and private comment. Both Faraday and Melbourne wanted it to die down but they also wished to make it appear that the reported initial conversation between them was imaginary. There were letters to the *Times* — Faraday himself wrote on 8 December denying that he had made any communication to the editor of *Frazier's Magazine* — and letters in other newspapers. Faraday's pension was granted on 24 December, he accepted, Melbourne thanked him for his 'willingness to contradict any injurious statements in the public prints',[45] and the case dropped from public notice. Peel's actions, however, had made the matter of scientific pensions a sensitive one for the Whigs.

The award of scientific pensions in 1835 had firmly established two points: (i) these pensions should be above politics, based on merit and (ii) they honoured science as well as the individual practitioner named. Peel had certainly gained by adopting these principles. Melbourne had been forced to recognize the powerful hold that scientific worth had on the popular mind and the usefulness of scientific patronage skilfully applied. On the whole Peel was the gainer, a fact made evident in a note sent Mrs. Somerville on 10 April by her old friend, the moderate Tory politician Sir Robert Harry Inglis. In 1829 when excitement over the question of Catholic emancipation was high, Inglis had unseated his fellow-Tory Peel as M.P. for the University of Oxford and had continued to hold that seat. On hearing of her pension he wrote her of his 'cordial gratification at the honour done by Robert Peel to himself in a late appointment connected with your name'.[46] In the same week the *Athenaeum* alluded 'with honest respect . . . to his [Peel's] liberal patronage of letters and science' and cited 'the merited pension settled upon Mrs. Somerville, the preferment of Milman . . . and an act of munificence to . . . [Crabbe's son]' as 'gracious and considerate acts'.[47] Peel set a standard by which to measure the disposal of Royal favour in the future.

In late May 1837, when the Somervilles had been for some time in serious financial difficulties, Lord Melbourne added another £100 a year to Mary Somerville's pension. This action partook more of charity than recognition of extraordinary merit, and was thus more in accord with Melbourne's instinctive views of the proper disposal of Royal favour than was Peel's gestures honouring scientific achievement. From evidence in the Somerville papers the chief figure in bringing about the increase was the Somervilles' old friend, Robert Ferguson of Raith (1771–1840; F.R.S., 1805), for many years Radical M.P. for Kirkcaldy. Mary Fairfax had known the Fergusons — an ancient and important Scottish family — since her childhood at Burntisland and had become an intimate friend of Ferguson's wife, Margaret Nisbet (1777–1855, once Countess of Elgin). Robert Ferguson had long been active on behalf of patronage for the Fairfax family, helped secure a baronetcy for Henry Fairfax (1790–1860) in 1836, and in that same year attempted to interest Melbourne in Mrs. Somerville's needs. Two notes from him in the Somerville papers not only illustrate the manner in which such favours were solicited but emphasize the absence of any scientific concern. The first, to Dr. Somerville, remarks:

. . . I have tried several times to see Lord Melbourne & missed him yesterday [4 August 1836] by a few minutes. But I had a note written to him, such as I would have stated to himself — that he has got — & in fact it is much better, for it is a better memorandum than a hurried conversation. If such a hint as I gave were to be given from some other quarters, I think it must produce the desired results. I stated the great satisfaction it had given, his bestowing £300 on certain literary characters — that it would be well to put the celebrated Mrs. Somerville in an equal footing — that it would be, without fail agreeable to *him,* to add the £100 — for the £200 was bestowed by *Peel.* [triple underlined] [48]

Melbourne's annoyance with Peel for his skillful use of civil pensions in 1835 to embarrass the Whigs was apparently well enough known to be included as an extra prod. But matters hung fire for some months. Charles Babbage must have interested himself, for after the additional pension was awarded Mary Somerville wrote him that Melbourne's secretary had told her 'of all the interest you have taken in me and how much I am indebted to you'. [49] In early May 1837 Ferguson instructed Mrs. Somerville to send him the information required so that he might transmit it to the Prime Minister. [50] On 27 May 1837 Melbourne's people wrote Ferguson that another £100 would be added to her civil pension, retroactive to 1 April 1837. [51] Once again Mary Somerville was lucky in the timing of the King's Warrant; [52] the official transaction was completed in late May and early June. On 20 June King William IV died. Had the pension not been already executed, a long delay would have been inevitable and the increase might well have been lost in the transition. A worsening economy discouraged new expenditures in 1837.

The whole transaction in 1837 was handled as a matter of party patronage, and Lord Melbourne's role was so remote that Mary Somerville in her memoirs ascribes the increase to Lord John Russell. [53] There was little public notice taken of the award at the time; a mere statement of the fact added to the report of a pension to Mary Russell Mitford appeared in the *Athenaeum.* [54] The additional £100 did, however, bring Mrs. Somerville's pension to the highest level given for scientific merit — on a par with those received by Airy, Faraday, Dalton, Brewster and Ivory. Charles Babbage in 1851 would declare, on the basis of then current reports, that £300 annually still represented 'the maximum of reward to science . . . [and] almost the minimum of reward for other services' — the military, the legal profession or the church. [55]

In the autumn of 1837 when the debate on the civil list customary on the accession of a new monarch took place. Mrs. Somerville was singled out for attack by Charles Buller, the Whig M.P. from Cornwall. In general the Whigs supported the bill, the Tories opposed it. Peel spoke strongly in favour of support for science and the recognition of scientists through pensions but declared he would vote against the measure because the sum it set aside was too small. Buller agreed with Peel. In arguing for a revision of the civil list, he cited Mary Somerville's pension — given as he emphasized by a Tory prime minister — as the worst misapplication of the intent of scientific pensions yet encountered. Hansard reports him as saying:

. . . Mrs. Somerville . . . he understood, was in perfectly easy circumstances, and her husband received large salaries from offices he held under the Government. However meritorious her researches might be, and no one wished less than he did to deprecate them, they were confined to the acquirements of branches of learning to which her sex had not aspired, and no one could undertake to say that they added anything to the stock of human knowledge or enlarged the bounds of science. Such a waste of money was to be the more deplored when he recollected the circumstances under which the grant had been made. . . .[56]

Two respected Whig members of the House, Sir Ronald Crauford Ferguson (1773–1841), younger brother of Robert Ferguson of Raith, and Francis T. Baring (1796–1866; F.R.S., 1849), immediately spoke in Mrs. Somerville's defence and against this personal attack.[57] She was in Scotland with the Robert Fergusons at the time and was shocked and distressed[58] to find her name appearing in the newspapers[59] in connection with such an unseemly furore. Her friends comforted her, the Civil List was finally approved at its previous level, and public talk gradually died away.

CHAPTER 9

'THE COMET', AN EXPERIMENT, AND A THIRD EDITION

'Our fate has changed, our gloomy prospects are brightening, and that fortune so long a stranger is beginning to smile upon us', Woronzow Greig wrote his mother in excitement at hearing of her pension. He went on to declare, in words that not only reveal her children's attitude toward their mother's science but express the sentiments felt by many of her circle and others, his admiration of Peel:

... that he in his sound judgement and upright generous feeling should be the first to appreciate the splendid talents and the indefatigable industry and the deep research displayed in the wonderful works which have come from your pen ... and come forward ... to wipe away the disgrace — too well merited — which attaches to the British nation, that it great and powerful tho' it be is perhaps the only one which has hitherto permitted the lovers of science — persons who sacrifice their time their health and their fortune in pursuit of knowledge the most useful & beneficial to the human race [—] to go without rewards, without thanks, even without notice — What a contrast does this generous conduct [by Peel] form with that of the late Administration the most powerful admirers and supporters of science will that ungentlemanlike scoundrel Brougham *ever dare* to look you in the face again? ... but enough of him he is not worth a thought, the blackguard. ....[1]

Congratulations came from family[2] and friends.[3] News of the pension quickly crossed the Channel, and Gerard Moll, who had defended English science from Babbage's charge of its decline but had lamented that the British government failed to honour itself in not honouring distinguished men of science,[4] expressed his delight in a letter from Utrecht.[5]

The Somervilles themselves could hardly believe their good fortune. Honours and happy tidings continued to pour in through the first half of 1835. The audience with Princess Victoria was followed by the long-hoped summons to the Queen's Drawing-Room.[6] William Somerville's name appeared for the first time on the House List of Proposed Visitors at the Royal Institution.[7] Arago wrote from Paris — too optimistically as it later turned out — that he had seen two or three chapters of Madame T. Meulien's translation into French of the *Connexion of the Sciences* and observed only slight inaccuracies; he thanked Mrs. Somerville for his copy of her new edition and assured her he would continue to supervise the translation.[8] Mary Somerville straightaway wrote her son that the translation would appear,[9] but in fact it was not published until 1839[10] and was neither well done nor successful. Arago had not been able to keep his promise.[11] Madame Meulien herself sent fulsome thanks in 1835 for Mrs. Somerville's gift of a copy of her work and her offers of advice.[12]

Election to the Royal Astronomical Society was followed by election to honorary membership in the Bristol Philosophical and Literary Society.[13] W. D. Conybeare was her chief sponsor, and his playful letter pointed out that she was now

. . . entitled to claim a nominal connection with many of the most distinguished Scientific Gentlemen of the country & though I believe we have not hitherto ventured to take the same liberty with any Lady, that must be attributed to the fact that so few of you have condescended to honor our pursuits by participation. We may trust however that many will be led by your example & then we shall hope that our lists may be more equally mingled unless indeed the superiority which you have indicated to your Sex, should ultimately induce you to exclude all of ours. . . .[14]

References to Mrs. Somerville's sex are strikingly absent from comment about her various honours at this date.

Of the sixteen British provincial scientific societies of the day, only Bristol made her an honorary member. Conybeare, who had met Mary Somerville in 1831 through Lyell and had since seen her at various social gatherings of the geologists in London, was a leading figure in the Bristol society. Although Cambridge admirers had not been able to persuade the Cambridge Philosophical Society to accept a woman in 1831–32, Bristol was more venturesome in 1835. Nowhere else in the provinces (save some years later in Newcastle) did Mary Somerville have an active advocate and nowhere else was she offered election. In London (where she had the admiration of the Royal Society and the special approbation of its reform element and the particular regard of the Royal Astronomical Society), in Dublin (where W. R. Hamilton was her special pleader), and in Geneva (where Prevost was her sponsor), recognition by leading scientific bodies was quickly forthcoming, but in her native Scotland the Royal Society of Edinburgh made no official gesture nor is there evidence that they were approached to do so. In the next three decades scientific societies on both sides of the Atlantic would confer memberships and diplomas on Mrs. Somerville — eighteen in all — but these honours came when she herself had reached such scientific eminence and achieved so great a scientific reputation as to be an asset adding glory to any group electing her.

The Bristol society promptly recognized their new member — and her gift of a copy of the second edition of her *Connexion of the Sciences* — by reviewing it in their *West of England Journal of Science and Literature*.[15] The unidentified author gives an interesting summary and appraisal of the work, making one point unnoticed by other reviews: the fact that Mrs. Somerville uses the word 'connexion' in two senses. In the first, she means those connections 'indicative of uniformity, or it may be of identity, in the laws in which . . . [physical] sciences are subject'. In the 'secondary and unavowed sense . . . [she means] subserviency'. One sentence ('Botany and Zoology are sciences *subservient* to Geology, number and quantity are sciences connected with Astronomy'[16]), together with the laudatory tone of the piece, suggests that the reviewer was probably Conybeare.

The first half of the year 1835 was for the Somervilles one of happiness, good fortune, and much sociability. A large number of the notes and letters of this period in the Somerville papers deal with parties in London and Chelsea, with invitations given or accepted. The two Somerville girls were 'out', although not formally part of the highest or most fashionable circles, such as those that gathered at Almack's.[17] They went with their mother — and sometimes Ada Byron came along also — to parties, receptions, other entertainments. Mrs. Somerville was well aware of the importance of bringing her daughters to the attention of eligible young men. Their future place in society and their future security depended, in her opinion and that of the public, on a fortunate match. They had no dowries and no expectations and, although intelligent, accomplished and ladylike, were not particularly pretty. They needed — and their mother perceived the fact clearly — as much help as possible.

She herself enjoyed many aspects of society. She disliked 'routs' and preferred smaller gatherings and good conversation but was always pleased to be among the fashionable. During these months there were numerous callers at Chelsea, old friends and new acquaintances. The liberal Scottish theologian, Thomas Chalmers (1780–1847) — a corresponding member of the Institute of France and much admired by many scientific men — brought his wife and daughter to meet Mrs. Somerville.[18] There continued to be, as there had been for almost a decade, a stream of foreign visitors, many of them scientific and medical men and their families. After the Somervilles' year in Paris and their introduction to its American colony, Americans began to find their way to Chelsea. The powerful Boston Unitarian circle gathered around W. E. Channing (1780–1842) and including many Harvard professors and other intellectuals were directed to Mrs. Somerville by Henry Bowditch and his father,[19] to whom in turn Mary Somerville sent Harriet Martineau when she toured the United States in 1834–36. Among the Americans who came to Chelsea in 1835 were Professor and Mrs. George Ticknor of Harvard. Ticknor (1791–1871) since his first transatlantic visit in 1815 had known many British and European scientific people and was anxious to meet the remarkable Mrs. Somerville.[20, 21] To Americans she was invariably cordial, regarding them with affectionate admiration and tolerant of the mannerisms and colloquialisms that sometimes amused, sometimes offended many English.[22] She claimed with pride a personal acquaintance with many distinguished American visitors to Europe, among them Washington Irving (1783–1859), James Fenimore Cooper (1789–1851), Joseph Tuckerman (1778–1840), Charles Wilkes (1798–1877), Benjamin Silliman (1779–1864) and James Dwight Dana (1813–1895) and correspondence with Matthew Fontaine Maury (1806–1873).[23]

Notwithstanding all the gaiety of these weeks, Mrs. Somerville did not abandon her scientific writing. She put together an essay on comets with special emphasis on Halley's Comet, and sent it off to John Murray for the

Quarterly Review. She had never tried her hand at such a piece and was anxious to get it right. Halley's Comet was scheduled to return in the autumn — its first appearance since 1759 — and was the subject of frequent notice in the press and on the lecture platform throughout the months of anticipation and immediately after its arrival. A long essay on it — now known to have been written by Dionysius Lardner — appeared in the April number of the *Edinburgh Review*.[24] Its rival, the *Quarterly Review*, could hardly let the astronomical event pass without an article of its own.

Mary Somerville seems to have written her piece on speculation, with no prior arrangements with Murray as to payment or date of publication. Her intention was to prepare readers, through an historical survey of comets and an exposition of cometary theory, for the phenomenon they would witness in the autumn. She must have submitted the essay in the late spring or early summer, for she wrote in July that she fancied the 'article on Comets would not do as the Review has appeared without it'.[25] Before the end of the month, however, proofsheets began to arrive. Along with them came a note from Murray, advising her that she 'need not hurry . . . about the Essay on Comets, as I will give . . . timely notice when it will be wanted, as soon as I know myself'.[26] Nevertheless, she worked on the sheets[27] and before the month was out could report that she 'was improving . . . the cometary paper very much and it is now in a state to afford me amusement rather than fatigue . . .'.[28] From that point onward, until its publication in the December number of the *Quarterly,* progress is well documented in the Somerville papers; three letters conveying the latest astronomical news attest the authority of the piece.

Returning to London on 2 September after ten days at Reading with the Richard Napiers, she found a note from J. G. Lockhart, delivered at Chelsea a week earlier, saying he was in town for a few days and wished the manuscript immediately, before returning shortly to his family vacationing in Boulogne. Working away 'as if auld nick himself had been at my elbow ready to give me a jab every time I looked up', Mary Somerville had the piece ready for him in a day and a half. At the very last moment a letter arrived from W. H. Smyth of Bedford, giving — as she wrote her daughter visiting in Scotland — 'a complete account of the comet which has kept its appointment within a week!!! . . . quite the thing for my review'.[29] Hard on the heels of Smyth's letter came a second note, this one from his wife — also an enthusiastic astronomical observer — urging Mrs. Somerville to come to Bedford the following evening (5 September) to view 'this interesting traveller before the Moon eclipses its faint light . . .',[30] a jaunt Mary Somerville would have liked but could not make.

Lockhart thought the piece 'much too long, so', Mrs. Somerville reported, 'I have cut off the Comet's tail and I am happy to say that the tail so cut off will make a review per se'.[31] Gerard Moll, on the other hand, dining with the Somervilles on the day the proof-sheets of the revised essay arrived, 'thought it a pity to leave out any [of the material]'.[32] Moll was in London after attending the B.A.A.S. meeting in Dublin, where he and Louis Agassiz

(1807–1873; F.R.S., 1838) had been the most distinguished foreign visitors present.[33] He willingly became Mrs. Somerville's temporary assistant in the project, undertaking one of the most important errands that her writing always required, a search of libraries of the learned societies. Dr. Somerville, who normally performed this service but who had spent the early part of the summer in Scotland, was ill after his return with a miserable cold that kept him in Chelsea. London itself was largely deserted in this season;[34] many of their scientific friends were abroad or still in Ireland. Mrs. Somerville, despite her recent election to the Royal Irish Academy, makes no reference in her surviving letters to any thought of going to Ireland, where Maria Edgeworth had long urged a visit. Moll's presence in London at this time, therefore, was doubly welcome.

The Dutch astronomer, however, was hard put to find the information he wished in the library of the Royal Astronomical Society. Despite the enthusiasm for science prevailing in Britain, and the increasing numbers of scientific publications, the libraries of the new specialist societies were always far from complete in their holdings and even what they owned was often inaccessible from lack of space. Not until 1830, when the Royal Society sold the Arundel manuscripts to the British Museum and then, under a bibliophile president, used the proceeds to buy English and foreign scientific books, was there any systematic effort to update the library of that august scientific body and to appoint a cataloguer who also served as part-time librarian. Dr. Somerville used these libraries on behalf of his wife, for she herself, with exquisite sensitivity, would not consider entering any such male sanctum.

Moll, who had so thoughtfully defended British science against charges of a decline, was shocked at the state of the library of the Astronomical Society. Written in English from the apartments of the Society on 12 September, his letter not only assesses objectively the tiny collection of books he found there in one cramped room[35] but gives a vivid idea of the kind of help Mrs. Somerville's consultants provided:

I have now rummaged in vain the library of this Society which, I am sorry to say, is still in its infancy. They have none of the books for which I was searching, no Pingré, no Littow, no Encke, nothing to our purpose. However, not to come to you with empty hands, I made an abstract of Delambre's relation of Messier's and Delisle's proceedings with regard to the last apparation [sic] of Halley's Comet.[36]

We rightly spelled the name of the Italian Astronomer, *Montanari*.[37]

I thought of searching Halley's poem on eclipses, which I might have had a chance of finding something about comets, but this book is also wanting.

I am sorry that I have not been able to execute your orders with better success, but my endeavours, at any rate [have not] been wanting.[38]

Although she herself was now — since 13 February 1835 — an Honorary Fellow of the Society, she would never have thought it proper to visit Society rooms without a special invitation and an escort.

Moll's praise of her essay encouraged her, for she was — as she wrote her son — 'anxious about the success of my first review as so much depends on it'.[39] Her manuscript was ready once again before the end of summer but once again publication was delayed.[40] Murray thought the piece might be in the January number,[41] then late in November he sent it to her 'with a note requesting it might be finished for the Press instanter'. She had 'to hunt right and left for [the latest] information about the comet & alter . . . [the paper] accordingly'.[42] By this time she and Dr. Somerville had themselves twice seen the comet at Sir James South's Camden Hill Observatory.

On 10 October they had been invited, through Charles Babbage, who was dining with Sir James on that day, to join the viewing party.[43] Mrs. Somerville's account of the evening reflects the mixture of science and snobbery customarily associated with such gatherings:

. . . We saw [the comet] beautifully and its nucleus is like the disc of a planet. I have no doubt of its solidity from its appearance. Its tail is beginning to grow. I assure you it was quite gay at Camden Hill and is so every clear night — There was Miss Fox [sister of Lord Holland], Lord and Lady Ashley [Melbourne's niece], Governor Gower & about 20 or 30 gentlemen. All the Fitzclarences [children of William IV] had been there a few nights before & Mr. Babbage has struck up a friendship with them. . . .[44]

A week later, after a second visit, she reported that going to Sir James South's 'is like going to a public place so great is the crowd, all the principal people in town which is rather amusing'.[45] The Smyths in Bedford repeatedly urged a visit to view 'the Fiery Monster'[46] and Captain Smyth regularly sent full and detailed accounts about his own observations and those he had from others, including Herschel's sightings of Halley's comet in South Africa on 1 September and of Encke's comet on 14 September.[47]

As time went on more and more recent information made its way into the comet essay. William Somerville reported that she now had the *Annuaire* and Pontécoulant's article on the event.[48] Young John Murray sent her, on its receipt, a letter from T. Jameson Torrie, nephew of Professor Robert Jameson of Edinburgh, dated 'Paris 4 November 1835' and reporting that 'the only novelty here is Arago's account of the Comet which I heard him give at the Institute last Monday week [27 October 1835]'. Arago's description of his observations from 1 through 23 October and his explanations for some of the phenomena he had seen were then outlined by Torrie. William Somerville made an extract[49] and the information, quickly inserted, appeared on page 221 of the printed essay to give an up-to-the-minute report of the latest ideas of Parisian astronomers about the comet.*

Francis Beaufort, on the day after the Somervilles dined with him on 18 November, sent Mrs. Somerville a copy of Herschel's new *Meteorological Observations*,[50] that had come to him. As soon as Lieutenant W. S. Strat-

---

*   In the Somerville papers there is a crude sketch which Mary Somerville, in notes made for her autobiography, identifies (under the heading 'Halley's Comet') as 'a rough sketch of this comet by Baron Humboldt drawn in our room at the hotel in Paris as seen by Arago & him at the observatory on the night of the 25th Octr 1835'.

ford's new ephemeris for Halley's comet observations came out in December, Stratford (1791–1853; F.R.S., 1832), now superintendent of the *Nautical Almanac*, sent her a copy.[51, 52] Notice of his accurate determination of the orbit appears on page 218 of her essay, which like her *Connexion of the Sciences* was rapidly becoming a report of the newest and most valuable work on comets.

Before the end of November Murray's printing schedule had again changed and Mrs. Somerville was called upon to work at top speed. Even so, the presses had to be stopped for the article. That emergency was fortunate, for, as she wrote her daughter:

> . . . now as it [the essay] ought to have come out last number, my account of the Comet was *pro*spective, and now I have been working hard to make it *retro*spective. accordingly I have got not without trouble the observations of Arago, Smyth, South &c and as good luck would have it this very day [28 November] arrived an immense packet from Admiral Greig containing the Russian observations of M. Struve perhaps the most valuable of all. The Admiral writes a very good kind letter[53] containing a great deal of scientific news [including some from the observatory at Dorpat]. . . .[54]

Greig had sent her an extract of a letter, dated 25 September 1835, he had just received from the astronomer F. G. W. Struve (1793–1864; F.R.S., 1827) at Dorpat. She immediately included Greig's information in the essay (pages 220–221), referring to him as a 'friend in St. Petersburgh' and writing her daughter that 'M. Struve's observations at Dorpat are by far the most curious that have been made and as yet unknown in this country'.[55]

Greig also enclosed a letter about another matter in which Mrs. Somerville had acted as intermediary. Richard Were Fox had designed an instrument for determining the direction and intensity of terrestrial magnetism, a question of growing interest at the time.[56] Hoping to get his device tried in Russia and a determination of the magnetic pole in Siberia, he appealed to Mrs. Somerville to call it to the attention of her brother-in-law Greig. She had done so, and Greig in turn referred it to the Imperial Russian Academy. The Permanent Secretary, P. Y. Fuss, reported after a time that the Academy could not underwrite the special trip the Siberian determination would involve, but that Academician A. T. Kuppfer (d. 1865) had agreed to make all the measurements in St. Petersburg and to compare the instrument — if Fox would send him one — with others 'qui passent pour les meilleurs dans ce genre'.[57] Fuss also answered some questions Greig had sent him on lunar observations. Mrs. Somerville in these years served repeatedly as a conduit for scientific interchange between Russia and the west. Her cordial relationship with her brother-in-law, his interest in science, his affection for Britain, and his high official position made her a simple and effective means of scientific exchange.

By 'writing from morning to night', Mrs. Somerville was able to send off her revised essay on 3 December,[58] and it appeared in the delayed December number of the *Quarterly*, ostensibly as a review of two accounts in German of Halley's Comet.[59]

Although excitement about the comet was beginning to die down, the piece seems to have been well received. J. B. Pentland wrote Dr. Somerville some three weeks after publication that he thought Arago would 'be much pleased with the article in the Quarterly on Comets. He is likely to publish something on the same subject'.[60] Biot he reported as being busy with Baily's new work on Flamsteed,[61] which was reviewed in the same issue of the *Quarterly*[62] that carried Mrs. Somerville's essay. She was particularly anxious that the article should succeed, for she hoped to do many other reviews on a regular basis. The Murrays encouraged her in this notion, for on 15 September John Murray Jr. had called at Chelsea to tell her that they wished her 'to undertake a series of reviews on scientific subjects'.[63] The news was most welcome, as the Somervilles were in the direst financial straits and Mary Somerville foresaw that only through her pen and strict domestic economy could they hope to survive.

When Peel had given her the pension in April, she for the first time since her second marriage felt that there was money in hand that was not immediately required for daily expenses. A note in the script characteristic of her old age and written on the copy of Peel's letter about her pension reads: 'No sooner received than blasted by trusting to the false & base under the mark of relationship and connection'.[64] Her reference is to the odious actions of a first cousin and contemporary, James Wemyss (died 1849), son of the minister of Burntisland and of the Somervilles' aunt-in-common, Christian Charters Wemyss. James Wemyss, a rakish but plausible fellow, borrowed large sums, lived extravagantly, ran up debts, then fled abroad, leaving several dependent spinster sisters behind in Edinburgh destitute and disgraced. William Somerville, without telling his wife, had some time earlier stood surety for their profligate cousin and was now responsible for the large sum borrowed by Wemyss from Forbes Bank.[65] Wemyss' failure had long been expected in the Scottish capital,[66] but to his London relations it came as an unforeseen and terrible blow. When, early in June 1835, Mary Somerville first heard the bad news, she was stunned.

For the third time since their marriage in 1812, Dr. Somerville and his wife faced a serious financial crisis brought about by the weakness of others and Somerville's unwise tendency to play the role of lord bountiful. In 1823, after the death of S. C. Somerville, they had raised money to avoid a posthumous bankruptcy in the family. In 1830 the fugitation of Somerville's Scottish man of business Henry Lowe had been costly to them. Now, two months after the award of a liberating pension, Mrs. Somerville learned what her husband and son had known for some weeks — that Dr. Somerville was in the most serious financial scrape yet. She had long been aware that their style of life often exceeded their income and that money was always scarce, but twenty-three years of marriage to William Somerville had taught her that he was loath to discuss finances or any other unpleasant matter. She knew that he, although 'not extravagant in his own person',[67] was inept and negligent in business matters, always taken by surprise when creditors became impatient, always ready to lend or borrow. She had no control, legally or personally, over his actions in such matters. Her only recourse in such emergencies was to strict domestic economy.

Somerville, without fortune but with tastes far in excess of his earned income, accustomed since childhood to dependency on the rich and powerful for place and position, was understandably convinced that association with influential people inevitably led into lucrative paths and was willing to risk much to maintain such associations. On many occasions he was rewarded: his appointment in London in 1816, his place at Chelsea in 1819, his position in London scientific and literary circles were all achieved through patronage and the friendly regard of others. His own considerable talents and his sunny willingness to be useful to friends and benefactors had — along with his wife's charm and ability — sustained this place. His intentions in money matters were always honourable but they were sometimes difficult to carry out. By borrowing here and there on occasion, by making judicious payments when absolutely necessary, and by keeping up an appearance of solvency while never crying poverty aloud, he had managed for almost two decades to survive among friends who were, for the most part, much more affluent than he could hope to be. The tie that brought him to the brink of financial ruin in 1823, in 1827, and again in 1835 was always the tie of blood, for he had borrowed money for his own brother, his sister-in-law, and finally for his cousin.

During all the financial turmoil in the Somerville family in the second half of 1835, William Somerville refused to discuss the Wemyss matter with his wife, 'unwilling [as always] to meet what . . . [was] unpleasant'.[68] She, however, soon after learning of the disaster, determined to cease being 'a passive spectator . . . [and to] take the management of affairs into' her own hands as far as possible. Her first thought was that the family should go abroad, where she believed they could live cheaply. At no time during this difficult period do her intimate letters reveal any grief at her own plight or the possibility of having to give up the scientific society she so enjoyed. Instead her thoughts are all of 'the blighted hopes of my poor girls'. She assumed blame for some of the difficulties the family now faced: 'Years have passed in a struggle with expenses which from our numerous acquaintance & proximity to London we have neither been able to meet or to fly from, an evil great in itself and aggravated by my unfortunate celebrity, as well as the necessity of introducing the girls'.[69] She rejoiced that they had recently 'furnished the drawing room [at Chelsea] and paid for it' but regretted that she had consented to the purchase of a carriage, which now must be sold.[70] Henry Warburton, 'to whom she laid open [in confidence] the whole state of affairs' advised her that she could not, as she had hoped, successfully 'apply for a continuation of my pension to my daughters in the event of my death . . . but entered so kindly into the whole business' that she was reassured by his friendliness.[71]

All thoughts of going abroad were soon abandoned. Instead Dr. Somerville and his daughters journeyed to Scotland, he to attend to business, the girls to have a little gaiety after the sad recent weeks. The ostensible reason for the trip — the Somervilles were determined to present an unruffled facade to the world — was the marriage of a Somerville niece in Scotland and the state of young Mary's health; she was under the care of Sir Charles M. Clarke, who treated female diseases.[72] Mary Somerville remained in Chelsea; the family's absence would enable her to put their household affairs and her own thoughts into more tranquil order. She herself was anxious to resume her writing, as she was 'convinced [that] the only way . . . is to pay the debt by annual installments by my pension'. She was, she declared, 'quite ready to begin to write again if I knew what, but it is not likely that I shall be as successful a second time', and she begged her son to find a subject for her, adding 'I must have something to do for I cannot be idle'.[73]

After her family had left for Scotland, she turned — as women often do in times of stress — to a thorough housecleaning at Chelsea. She instituted strict household economy and a simpler, quieter style of life. She refused invitations to balls but dined out with friends and received callers.[74] Murray kept her supplied with new books and these, with her painting and work on the comet essay, kept her occupied. Her life was dull only in comparison with her daily rounds in the first half of the year; the letters of these later weeks are filled with names of callers and accounts of small gatherings of friends. Occasional passages reflect her own amusement at — and enjoyment of — her fame even in these unhappy days. When a letter from an old friend of Dr. Somerville — a gentleman unknown to her — arrived offering a visit, she replied that the doctor was away and would regret missing him but that she would be glad to make his acquaintance. The gentleman sent his wife in his stead, but she would not come in, merely leaving her card. 'I fancy', Mrs. Somerville wrote her daughter, 'she was afraid of the claws of the lioness, she need not, for they have been clipped of late'.[75]

The 16 August issue of Leigh Hunt's *Examiner*[76] carried an extract from the new edition of Lydia Tompkins's *Thoughts on the Ladies of the Aristocracy*[77] which gave a breathlessly exaggerated account of Mrs. Somerville's struggles as a young girl to learn mathematics. She was amused by the 'puff',[78] Dr. Somerville flattered.[79]

By mid-August she had recovered her spirits to the point of going for ten days to stay with her friends the Richard Napiers at Reading. Shortly before she left she had been cheered by the £360 that came from Murray's discounted bill for royalties, 'a tolerable haul for my second edition and very seasonable' as she described it.[80] She hoped too the comet piece would bring another £50[81] and rejoiced when young John Murray mentioned 'a series of reviews on scientific subjects' since through them she might 'be able to pay off the Forbes [debt] sooner than we thought'.[82]

William Somerville returned to Chelsea on 5 September, fatigued and ill with a cold. The two girls had remained in Scotland to make a long visit to family and friends. They intended to spent part of October in Edinburgh visiting the Jeffreys and the Horners, go on to Raith to stay with the Robert Fergusons, and then to Limekilns, the seat of old family friends, the Grahams. Woronzow Greig in mid-October 1835 proposed to the younger Graham daughter Agnes and was accepted. The match was already in the air when Martha and Mary Somerville travelled to Glasgow and all were delighted that they would have opportunity to know their future sister-in-law better. Not until early in January 1836 would the two girls be back at Chelsea, after six months in Scotland.

Without her daughters and worried about the future, Mary Somerville turned once more to experimental science. She began a simple investigation of the permeability of various bodies to the chemical rays of the sun,[83] a study she could carry out in her Chelsea garden in this hot early autumn, using various handy filtering media. Neither she nor her contemporaries recognized that she was in fact employing a kind of primitive photography in this study. Her interest was not in producing an image on a sheet of paper coated with silver chloride, using light, but rather in observing the degree of discoloration produced in the paper so treated by the passage of the 'chemical rays' of the sun through various media — clear and coloured glass, clear and coloured mica, jewels such as emeralds, garnets, beryls, tourmalines and the like, and rock crystal laid in turn on the paper and exposed to sunlight. She apparently consulted Michael Faraday about the experiment, for on 12 October he wrote her:

I have been making some experiments with the papers but do not succeed in obtaining so good & regular a result as I wished & believed I might obtain.

In the first place the precipitates made upon the paper are not so sensible or regular as that first formed & washed & applied in the usual way, the excess of the muriate or nitrate used & the resulting salt formed interfering with action of light by retarding more or less the change and that in an irregular manner. Chloride produced on the paper is therefore nothing like so regular in its change as chloride previously precipitated & well washed.

In the next place I do not find that I can lay a more regular coat of the substance in the method I mentioned than by using the moist precipitated chloride & a camel hair pencil.

I suspect your chloride is a good deal [illegible]. I will therefore precipitate & wash some and send it to you in the moist state. Allow me to suggest that when you refer to and apply it to paper for your experiments you do so in a dark place, or by candlelight only & thus you may keep it for a long time in good condition.

I send also Biot's report for your inspection.[84]

What she set out to determine was whether the chemical rays of the sun, which were known to blacken silver and fade vegetable colours, displayed activity analogous to that shown by light rays and the calorific rays passing through clear and coloured glass. The accepted view of the solar spectrum at that time was that it consisted of five overlapping spectra, three of them — red, yellow and blue — being visible. At one end of the visible part was the calorific rays spectrum, at the other the chemical rays spectrum. Arago's protégé Macedonio Melloni had found in his investigations of radiant heat 'that although the colouring matter of glass diminishes its power of transmitting heat . . . [only] green glass possesses the peculiar property of transmitting the least refrangible calorific rays, and stopping those that are more refrangible'.[85] Mrs. Somerville was familiar with Melloni's work — she had included it at increasing length in successive editions of her *Connexion of the Physical Sciences* — and Faraday had been so impressed with it that he persuaded the Royal Society to give the Italian its Rumford Medal in 1834 and in the spring of 1835 had devoted his own Friday discourses to Melloni's recent discoveries.[86] In essence Mary Somerville proposed to make the same sort of study of the chemical rays of the solar spectrum.

Her investigation exhibited her usual careful experimental design. For the most part she used materials easy to come by — bits of coloured glass, jewels, rock salt, and the like. She neatly and simply executed a series of experiments that considered a variety of parameters: different media, different colours and shades of colours and different shapes in these media, different times of exposure to sunlight, different degrees of solar heat, and so on. She carefully maintained controls so that she might interpret her results. Distinguishing degrees of activity simply by ocular measurement of discoloration of silver chloride was difficult, but she found that rock salt transmitted the greatest number of chemical rays, violet and blue glass fewer, and green glass the least number.

In the Somerville Collection are two rough drafts in Mary Somerville's hand — both fragmentary and undated — of a letter to 'My dear Sir' — Arago — describing her experiment and her results, 'new as far as I know'. She informs her correspondent that she would be 'much gratified' if he thought them 'worthy of the notice of the Royal Academy of Science' but asks that he 'have the kindness to verify them before presenting them to that distinguished Society'.[87] Arago at the 21 December 1835 meeting of the Académie des Sciences presented an extract of Mrs. Somerville's letter. His report was later published in *Comptes rendus,* and constitutes Mary Somerville's second published paper. In his remarks to the Physics Section of the Académie about the paper, Arago pointed out that he had abstained — not wishing to infringe on his colleague Melloni's work — from mentioning some results he himself had obtained on the absorption or interception of the chemical rays but that the same reserve could not be demanded of Mrs. Somerville. 'Je ne vois donc point de raison', he added, 'pour refuser aux intéressantes expériences d'une personne si éminent distinguée, toute la publicité des séances de l'Académie et du Compte rendu'.[88]

Mrs. Somerville's reasons for sending her account to Arago are obvious. She had discussed the phenomena of the solar spectrum with him for almost two decades; she knew he was interested in the sort of work that Melloni — and now she — was doing; she recognized that with him there would be no difficulties about her acceptance of the undulatory theory; John Herschel, who would have been the most likely choice for presenting the paper to the Royal Society, was still at the Cape of Good Hope. In fact, Arago served admirably as a means of transmission of her work to the world of science.

This experimental work provided a welcome change from writing. Although she now felt pressed to earn as much as possible through her pen,[89] the despair she felt in August made authorship a chore.[90] To her son in Scotland she wrote, 'I wish that you would concoct something for me to do that would be ornamental and useful . . . I am weary of *science* I have had so much of it, and should like to try my hand in some other branch if I could hit it'.[91] By September, however, she was delighted to hear that Murray wanted a series of reviews and by October reported with pleasure that 'John Murray tells me they will keep me busy which is exactly what I want. If I could write two or three articles annually it would be worthwhile'.[92] In mid-December she declared, 'I intend to write an article on meteors for the next number of the Quarterly if they will take it . . .'.[93]

Richard Napier from Reading sent a suggestion that she undertake a translation of 'Pontécoulant's Notice [sur la comète de Halley]' that had just come out in Paris. He thought it would be a service to society and was glowingly confident that such a work, under her name and sold at a low price, would be in great demand at the 'Scotch, English & Irish Universities . . . Philosophical & Mechanics Institutes and the Reading Clubs . . .'.[94] He also urged a translation of Laplace's *Système du Monde,* with explanatory notes. Neither was undertaken.

What did require to be done was the preparation of a third edition of *On the Connexion of the Physical Sciences.* After the quick sale of the second edition Murray was convinced that a third one would be profitable if brought out at the right time. Mary Somerville seems to have begun its preparation almost routinely. In March 1835 she had a kind letter from her friend the historian Henry Hallam (1777–1859; F.R.S., 1821)[95] pointing out some inaccuracies in her passage in the second edition about the length of the ancient Egyptian civil year. He assured her that he offered this information solely because he was confident another edition would soon be wanted. Mrs. Somerville thanked him, pointing out that she had been misled by some paragraphs in Biot's *Astronomy,*[96] and asking for any other observations 'you may favor me with as I have reason to believe another edition will soon be called for'.[97]

In June William Somerville asked his colleague at the Geological Society, W. H. Fitton, to read some of the sections relating to his speciality. Fitton replied that he would be delighted to do so, after he had finished some tasks demanded before August. He ended his letter with the flattering words, '. . . in reading the Book I had repeatedly occasion to observe that it was so well written — as to require *very* little alteration indeed!'.[98]

These letters and one other, from Babbage, are the only ones, apart from the Murray and family correspondence, among the Somerville papers that refer to the third edition. This paucity is in sharp contrast to the heavy correspondence with consultants about the first and second editions. Three reasons, other than possible random loss, suggest themselves as possible explanations for this diminution in scientific correspondence. First, William Somerville, who made many of his wife's scientific enquiries and contacts, was away in Scotland for six weeks in July and August 1835. Second, London in that burningly hot summer was largely deserted by their scientific friends, many of them staying on in Ireland after the B.A.A.S. meeting and others away on the Continent or in the country. Finally, the Somervilles themselves were in a state of great despondency, their attention given to family, rather than scientific, matters.

The success of the first two editions, however, seems to have given Mary Somerville new confidence in her ability to decide what to include and what to remove from the work. In September she wrote her daughter in Scotland that she was busy with getting '. . . my third edition ready for the press . . . I thought it finished, and now find new matter to add or old to improve'.[99] In October she appealed to Charles Babbage for information about the reports Professor Baden Powell had made at the Dublin meeting of his new researches on radiant heat — one of the liveliest topics of the day — and on experimental confirmation of some recent mathematical findings in undulatory theory. Babbage in one of his characteristically short notes told where 'Profr. Powells paper on Dispersion read "popularly" at the British Assn.' could be found.[100]

By 20 November the book was actually printing.[101] Dr. and Mrs. Somerville were invited to spend Christmas with Lord and Lady King at their country seat, Ockham Park, in Surrey. Ada Byron and Woronzow Greig's good friend William King, who had succeeded to the title in June 1833, had married in July 1835 and cordially wished the Somervilles to be among their first houseguests. Lady Lovelace wrote, 'You can spend as much of the day as you like in company with your Proof Sheets, which can easily be transferred backwards and forwards to Town'.[102] Possible profits spurred the author on. She hoped, she wrote her daughter in Scotland, 'in the course of 6 months to get a hundred or two for my third edition' and with that and her pension to pay off in three years the money they had been forced to borrow.[103] At Ockham Park with the Kings, she kept smartly at work but the printers were maddeningly slow.[104]

The official publication date was set for 1 April. A few early copies of the work appear to have reached the Somervilles around that time for the first was dispatched to Sir Robert Peel. He thanked the author for her favour in a letter dated 30 March in which he declared himself

. . . sincerely gratified by this new Proof that your Labours in the great cause of Science are properly appreciated, and will secure to you that Fame, which, next to the Consciousness of having applied your Talents to worthy objects, will be your chief Reward.[105]

On the other hand, Mrs. Somerville appears to have been putting the finishing touches to the work at this time, for she wrote her son on 6 April, 'I am busy with my Index which is enough to distract a stronger head than mine and to wear out a much sweeter disposition'.[106] From letters in the Somerville papers, only Peel and the writer-librarian at Holland House, John Allen (1771–1843)[107] are known to have received presentation copies.

The third edition when completed resembled the second in its inclusion of new material and improved presentation of the old. Melloni's experimental work described in the hastily-added 'Supplement' to the second edition was now woven into the text. Professor Baden Powell's researches[108] on light and radiant heat the previous year are mentioned. The most striking change is in the section on comets. Mrs. Somerville, armed with the material she had assembled for the *Quarterly Review* piece, vastly enlarged the factual information in this section. The pattern of the book remained the same, plates were still included and explanatory notes at the end continued.

Murray printed 2000 copies, a thousand fewer than the number of the second edition but equal to the first. His accounting statement[109] shows eleven copies deposited at Stationers Hall, as required by law, two copies sent to reviews, and five delivered to the author. Costs were £293.17.3, the smallest yet.

On 1 March 1836, over a month before the book was out, Mary Somerville wrote her son that pre-publication sales were brisk:

. . . The best news I have to tell you is that John Murray sold *1800* copies of my third edition at his bookseller dinner and there were only 2000 copies printed so it is nearly sold off at once. I think myself a wonderfully lucky dog.[110]

The *Athenaeum* on 5 March reported the event in some detail.[111] Mrs. Rundell's *Domestic Cookery* still stood at the top of Murray's list with sales of 6000 copies of its 59th edition, but Mary Somerville with the reported disposal of 2000 copies outdistanced every other scientific writer by 500 copies or more and surpassed all his other offerings save Byron, Crabbe, Boswell, and of course, Mrs. Rundell. This heavy demand meant that another edition would be called for when Murray judged the time right.

Two months before publication, in February 1836, Murray began to list the new edition among his offerings.[112] In April, after virtually the whole edition was sold to the booksellers, he gave the work an advertisement all to itself, first in the *Literary Gazette*,[113] then in the *Athenaeum*.[114] This puff was for the benefit of the booksellers. Once again he quoted the extract from Holland's 1834 notice in the *Quarterly Review*, words he had used in pushing both the first and second editions.

No other edition would sell so quickly or so easily. In October Murray once more called attention to the work, listing it first among nine new scientific and medical works from his house. *On the Connexion of the Sciences* was followed by Faraday's *Chemical Manipulation* and the fourth edition of Lyell's *Principles of Geology*.[115] Mrs. Somerville's work, by the time of its third edition, had established itself as a classic and a scientific bestseller.

CHAPTER 10

THE LAST LONDON YEARS

1. A New Pattern of Existence, 1836

For the next fifteen to twenty years, from mid-1835, the pattern of the Somervilles' existence would largely be shaped by financial circumstances. Each August until the Forbes debt was finally discharged was a time of anxiety and crisis, for payment was due in that month. To the outside world, unaware of these pressures, Mary Somerville appeared an object of serene good fortune. She reported to her daughter with wry amusement that late in 1835 some of the Chelsea pensioners had been heard to say:

. . . "How easy it is to gain two hundred a year & how lucky to be called clever: the government had little to do with the public money when they gave two hundred a year to Mrs. Somerville for writing a book that nobody can understand."[1]

The strict economy that Wemyss' dishonourable conduct forced upon the Somerville ladies was not so devastating to them as was the loss of that warm glow of security and minor affluence that had for a moment engulfed them. In the future Mary Somerville's pension became in their minds the core of the family's financial structure rather than merely a happy surplus. To others their life appeared one of dignified simplicity, arrived at through choice. They grieved in private over the loss of a carriage, the absence of a second London season for the girls, the careful curbing of even the smallest spontaneous expenditure.

Dr. and Mrs. Somerville continued to see their friends, to enjoy scientific company and, in fact, to be much in London society, but they themselves were always conscious of their changed circumstances. Early in 1836 Mary Somerville finished her second essay for the *Quarterly Review*, one on meteors for which she hoped to get £30 to £50.[2] She waited anxiously for its appearance[3] but Lockhart never used it. No reason is found among her papers; there are only fragmentary sheets on meteors in William Somerville's hand.[4] He himself in the spring undertook the role he best enjoyed, that of intermediary. In this instance he acted to heal the breach existing since the publication of the *Reflections on the Decline of Science* between John Murray and Charles Babbage. In Babbage's investigations of the affairs of the Royal Society he had come upon material which led him to charge that the transaction entered into by the Council and John Murray in 1826 to print Davy's presidential discourses had resulted in needless loss to the Society and unwarranted profit to Murray.[5] This charge Murray justifiably resented.

Babbage in his autobiography explains that while on a visit to Wimbledon Park, the residence of the Duke of Somerset (1775–1855; F.R.S., 1797):

. . . Dr. and Mrs. Somerville came down to spend the day. Dr. Somerville mentioned a very pleasant dinner he had had with Mr. John Murray of Albemarle Street, and also a conversation relating to my book, "On the Decline of Science in England". Mr. Murray felt hurt at a remark I had made on himself whilst criticizing a then unexplained job of Sir Humphry Davy's. Dr. Somerville assured Mr. Murray that he knew me intimately, and that if I were convinced that I had done him an injustice, nobody would be more ready to repair it. . . .[6]

Somerville passed the message on to Murray, who sent Babbage a copy of his accounts, figures that showed the publisher had lost rather than profited by the transaction. Babbage after re-reading the offending passage and studying the accounts wrote Somerville that his charge had been inaccurate and asked him to ascertain what amends would be acceptable to Murray so that Babbage might 'set myself right in his good opinion'.[7] A fortnight later Babbage wrote again, offering to prepare 'a short supplement [to his *Reflections*] stating fully the facts and acquitting Mr. Murray of any of the slightest blame . . .' or to insert an account of the true facts in the *Annals of Philosophy* or to undertake any other means Murray suggested.[8]

Murray and Babbage did become friends again and in the following year Murray published Babbage's *Ninth Bridgewater Treatise*,[9] written to redress what Babbage considered a shortcoming in the eight commissioned treatises, namely insufficient proof of the Omniscience of God. Babbage discussed the work at length with Mary Somerville[10] and made her a present of an early copy.[11] He had now become her 'favourite of all the scientific'.[12] In July 1836 when Baron F. P. Charles Dupin came to England for the sixth meeting of the British Association, the Somervilles asked Babbage to dine with him at Chelsea.[13] Babbage's excited acceptance carried a postscript: 'I have a kind of vision of a possible developing machine'.[14] He was still confident his calculating machine would be completed. Both Babbage and Dupin — also interested in machinery used in manufacturing — took active parts in the Statistics Section at the Bristol meeting.[15]

Once more Mary Somerville was invited to attend[16] and once more declined. Fatigued and worn by the crisis of 1835, she felt far from well and the severe headaches from which she had always suffered recurred with increasing frequency and pain. A change of air was strongly recommended. A short stay in Germany would be cheap and worthwhile. She wrote her son, 'I hope to lay in a stock of health for the winter to enable me to make out a fourth edition which I am secretly desired to prepare'.[17] Murray had seen how quickly the third had gone and wanted to be ready with a fourth while not jeopardizing sales of the current one.

Their stay in Germany was short, pleasant, and — save for a visit to

Werner's mineral collection at Freiburg in the Black Forest[18] — non-scientific. Once back in London, however, Mrs. Somerville resumed her place in scientific society with enthusiasm. She had reached the point where she —who had long been the petitioner for help and information — had become a great figure to whom homage was paid and introduction sought. Strangers and friends sent her letters and offprints and books and asked her advice. J. D. Forbes,[19] for example, hurrying 'through Town [on his way] to Bristol' for the B.A.A.S. meeting, left for her a copy of his second paper on polarization of heat, lest she miss seeing it; his researches immediately made their way into her fourth edition. New names, such as that of Lagrange's nephew, the mathematician G. A. Plana (1781–1864; F.R.S., 1827)[20] and that of the Belgian physicist J. A. F. Plateau (1801–1883; F.R.S., 1870),[21] began to appear among her letters. Two of her new correspondents, the travelers John Gardner Wilkinson (1797–1875; F.R.S., 1834)[22] and Mountstuart Elphinstone (1779–1859),[23] would be of assistance in giving her information for the work on physical geography she was already beginning to plan.

Another new acquaintance was the American naval officer Lieutenant Charles Wilkes, who was in Europe buying instruments and gathering information in preparation for a great surveying expedition to the Pacific and the South Seas.[24] He and Francis Baily had become friends and Baily introduced Wilkes into his special scientific circle, which included the Somervilles. Wilkes and Mary Somerville later corresponded about his exploration in the Pacific area.[25] There were other American visitors in this year, one of them being Mrs. Eliza W. R. Farrar (1791–1870) of Boston,[26] whom the Somerville ladies had wanted to meet since reading her child's life of their hero Lafayette.[27]

Mary Somerville's celebrity continued to grow. In December 1836 she sat for William Wyon (1795–1851),[28] chief engraver at the Royal Mint, who wished to add her profile to his collection of other notables, including the King. She dithered about how to do her hair — her daughters were in Scotland and not at hand to advise[29] — but considered Wyon's drawing of her for the medallion 'excellent'[30] and liked the finished medal. By the end of the year the Somervilles had settled into a pattern of existence that would obtain for the remainder of their residence in England.

2. The Fourth Edition of the *Connexion of the Sciences*

By December 1836 sheets for the fourth edition of *On the Connexion of the Physical Sciences* were arriving at Chelsea, and during the deep snows of that mid-winter Mary Somerville worked away at them,[31] 'by no means amusing occupation'[32] but one relieved by William Somerville's reading aloud to her as she worked.[33] On 1 January the venerable Governor of Chelsea Hospital, Field Marshal Sir Samuel Hulse (1747–1837), died. For all practical purposes he was the only governor the Somervilles had ever known; his great virtue in their eyes was that he 'knew the art of governing to consist in the principle of letting alone . . .'.[34] In 1832, after the death of the Surgeon at Chelsea, Sir Everard Home, William Somerville had been given Home's duties but not his stipend.[35] Now they were anxious lest there be further unfortunate changes such as the King's putting a strong Tory sympathizer in the place.[36] The Somervilles were now clearly identified as staunch Whigs, perhaps even radicals. In fact the appointment was long delayed by a series of unexpected deaths and was not made until 1839.

Three pieces of good fortune, however, brightened 1837. The year opened with the appointment of Dr. James Craig Somerville to a second inspector-ship of anatomy. Since the passage of Warburton's Anatomy Bill in 1832, James Somerville had been Inspector of Anatomy for Middlesex and he was now given the inspectorship also of Scotland, with an additional £400 a year.[37] In May Mrs. Somerville's pension was increased by £100 annually. In Octo-ber Woronzow Greig married his Scottish sweetheart, Dr. and Mrs. Somer-ville went to Scotland for the wedding, and Mary Somerville remained for nearly four months with her friends, the Robert Fergusons of Raith.

Although she herself now considered her London life much curtailed, Mary Somerville's name continues to appear in journal accounts and letters written by various celebrities in the English capital in 1835, 1836 and 1837. In April 1837, for example, the American scientist Joseph Henry (1797–1879) of Princeton College in New Jersey, U.S.A., abroad for the first time and intent on acquiring the latest philosophical information, books and apparatus and on becoming acquainted with as many foreign scientists as possible, wrote at some length of his meeting Mrs. Somerville and later visiting her in Chelsea. She, undoubtedly on Faraday's advice, had given early recognition to Henry's 1831 work on electromagnetism[38] by mentioning his giant electromagnet at Albany in the first edition of her *Connexion of the Physical Sciences*.[39] This short notice she maintained unchanged — despite Henry's subsequent re-moval to Princeton and then to the Smithsonian Institution — in later edi-tions,[40] a practice quite contrary to her usual updating and elaboration of text. Earlier in Philadelphia — in April 1834, two months after its publication in London — Henry had seen a copy of her *Connexion* in the study of his friend Isaac Lea (1792–1886), the malacologist and publisher, and had written in private notes, 'Was pleased to find she made mention of my experiments in magnetism'.[41]

Henry was anxious to meet her when he arrived in London in the spring of 1837 and soon had opportunity to do so. He and his fellow-American, Professor Alexander Dallas Bache (1806–1867; F.R.S., 1860) of Philadel-phia, a great-grandson of Benjamin Franklin and himself a well-regarded natural philsopher, were presented to the celebrated lady on 8 April at one of Babbage's evening parties. Bache in a letter written the next day described her as being 'without affectations as much at home in the occupations & amusements of the ladies as in the study of the men & withal too amiable to avail jealousy'.[42] Ten days later he and Henry, accompanied by Mrs. Fara-day, visited the Somervilles in Chelsea. Henry, in a long journal entry for 19 April 1837 declared that

Mrs S appears about 45 [she was in fact 56] is a very unassuming and interesting person. She is much caressed by the nobility and gentry. While we were there several persons of these classes came to pay their respects. The room into which we were shewn was hung around with Pictures of various masters and amoung [*sic*] the number some very beautiful landscapes in oil which we were surprised to learn were the production of this most talented individual. She is also well skilled in Music and performs admirably it is said on several instruments.[43]

A much garbled personal history, clearly derived from hearsay rather than directly from the lady, follows.[44] Henry also confuses edition numbers, reporting that, 'She is now preparing for the press the fifth edition of her work on the connection of the sciences'.[45] It was a copy of the fifth edition that he later obtained for his own library,[46] but in the spring of 1837 Mary Somerville was busy with her fourth edition.

Murray records show its publication date to be September 1837,[47] but the title was not included in the 'List of New Books' in the *Literary Gazette*[48] until 18 November and not at all in the *Athenaeum*. On 23 December Murray listed the work for the first time, giving it third place in his list of 26 titles.[49] The Murray account statements for 1837–38 are not among the Somerville papers, but a sheet in the hand of a Somerville daughter[50] — and confirmed by Murray records[51] — shows 2000 copies printed and royalties of £173.12.1 received from the edition. The price was kept at 10s.6d.

The work was again dedicated 'To the Queen', the inscription being identical to the one used in the past and no reference being made to the fact that since 20 June 1837 Victoria rather than Adelaide had been on the throne. As with the third edition, the 'Preface to the Fourth Edition' read simply, 'In order to keep pace with the progress of discovery in various branches of the Physical Sciences, this book has again been carefully revised'.[52]

Its table of contents, however, showed only five changes from the third edition, two of them being merely differing identification of material already present.[53] The other three changes are more interesting. The first is the inclusion of discussion of the ideas of Professor Ottaviano F. Mossotti of Turin (1791–1863), who hypothesized that one general law accounted for all the different forces in matter. He suggested that

. . . particles of matter attract one another [by gravitation] when separated by sensible distances . . . repel each other when they are inappreciably near . . . [and] that there might be some intermediate distance at which the particles might neither attract nor repel one another, but remain balanced . . . in stable equilibrium . . . electricity . . . [being] the agent which binds the particles of matter together.[54]

Faraday, receiving a copy of this memoir from Turin in December 1836 — at the time Mary Somerville was correcting her sheets — was so 'struck with it' that he sent it to Whewell, asking if the mathematics were right,[55] and to the printer and naturalist Richard Taylor (1781–1858) for inclusion in his new volume of foreign scientific memoirs.[56] Mossotti had no experimental evidence but his theory, if correct, accounted for gravitation and cohesion and supported a presumption of a universal ether. Faraday, who was on the verge of proving that electrostatic induction was not action at a distance, devoted his four Friday discourses in 1837 to Mossotti's work.[57]

Mary Somerville, who was frequently in touch with Faraday,[58] seems to have caught his enthusiasm, as did Ada King. Before the Somervilles went to stay with the Kings at Ockham in early February 1837, Lady Lovelace wrote her: 'I am full of curiosity about Mossotti's new theory, which I expect you to make intelligible to me'.[59] In May Mrs. Somerville — who had already sent a short account of the theory to Prince Koslofski — wrote him of its reception in England and added other scientific comment:

. . . I have not heard how it [Mossotti's theory] has been received by the Continental philosophers but it is highly approved of in London and at Cambridge. The analysis is published in the 3rd number of a new periodical work of high promise which is just begun and is to be continued every three months, it is Taylor's Scientific Memoirs . . . price 6 shillings. I am very pleased with the first three numbers but I fear it is too analytical to have an extensive sale or to succeed at all. Professor Whewell of Cambridge has just published a work in 3 vol. on the history of the progress of the human mind with regard to science [Whewell's *History of the Inductive Sciences* (1837)]. I have not seen it nor has any one had time to read it yet so recent is its appearance, but as he has a great name much is expected. To read our advertisements you would imagine that we are the most scientific nation in the world but a very excellent book is a rare thing. I must say, however that there is much activity though it is generally directed to the elementary branches, education being the great object at present among all classes. . . .[60]

Mrs. Somerville's new edition continued her endeavours to educate readers in the latest scientific ideas.

The second change in her *Connexion of the Sciences* was inclusion of an account of the work on circular polarization of calorific rays of the sun done recently by Forbes and by Melloni[61] and of her own 'experiments on the transmission of the chemical rays through transparent media', which she identifies as being made 'during the preceding autumn [by] the author . . . .'.[62] A report of 'M. Arago's observations in an Artesian well now boring in Paris'[63] supplies data on thermal springs. In discussing terrestrial magnetism and the existence of electric currents 'in metalliferous veins' she refers to the experiments of R. W. Fox in Cornish mines and adds, 'Even since the last edition of this book was published, Mr. Fox has obtained additional proof of the activity of electro-magnetism under the earth's surface'.[64] Fox had written her in some detail of this proof in late April 1836.[65]

In this fashion, from letters, reports, papers and books, Mary Somerville would continue over the next three decades to bring out revised editions of the book. After the fourth edition, however, there was rarely opportunity for direct conversation with experimentalists and theoreticians, one of the factors that made her first four editions so notable. By 1836, the skeleton and much of the flesh of the work was firmly established. Future changes — and they would be steady — came about through discarding old ideas and inserting new ones. By 1836 also her text was so suitably organized and broad as to accommodate the changes needed to keep it up-to-date. The difference in tone in the correspondence about the first and second editions and that about the third and fourth is striking, and it reflects a new assurance on her part. Consultants were not now asked to read each sheet, advise and correct. Rather they supplied new information, new ideas and their own valuable assessments of recent work and of the directions science was taking. Mary Somerville, preparing her third and fourth editions, is no longer unsure of how to proceed, no longer uncertain of her own powers of judgement. Her success had convinced her that she could measure the value of scientific material and present an accurate account of it. In March 1837, working away at the fourth edition, the new confidence in her own powers was confirmed when 'a gentleman at Mr. Babbages last night [told me] that he had given two guineas for an old copy of my first book . . .'.[66] For a moment she again thought of a second edition of *The Mechanism* but none was ever to be.

Her *Connexion* had a wide readership. Her scientific audience had been apparent for some time and she often learned of other readers. On the way to Scotland to her son's marriage she encountered one of her provincial readers. After a day and night in a carriage, one of the passengers asked Dr. Somerville 'if that lady sitting opposite to him was Mrs. Somerville author of the excellent book on the Connexion of the Physical Sciences whose bust he had seen and admired at Chantrey's?'[67] Their fellow-passenger was a self-made mining engineer, Thomas Sopwith of Newcastle-upon-Tyne (1802–1879; F.R.S., 1845). He insisted that the Somervilles stop with him when they reached Newcastle, proudly showed them his copy of her second edition, and became a lifelong friend. *On the Connexion of the Sciences* also made its way into superior school-rooms. The three children of R. W. Fox, aged 16 to 19, spent an hour each day in 1835 doing 'Italian, Somerville, Mathematics and Arithmetic'.[68] Whewell, on the other hand, asked in 1854 to recommend a book treating of light for a niece, advised her that 'Mrs. Somerville's "Connexion . . .", although written by a lady, is, I think, the hardest of all [popular science books]'.[69]

3. A Scientific Intermediary

Mary Somerville's stay in Scotland in 1837 was the longest visit she had paid her native land since coming to London in 1816. She was made much of by family and by Edinburgh society and enjoyed the weeks she spent with the Robert Fergusons at their three large Scottish estates but she returned to London with relief. This metropolis was now her home, the place and the society she enjoyed best. While in Scotland she — recognizing all that London and her friends there had done for her — wrote her daughters that had she lived in Scotland

. . . I never should have written a word and you two w^d have been worried to death on

Stuper [*sic*] like the rest, so all is for the best. Jedburgh my birthplace I saw without pleasure and left without regret, yet the vale is most beautiful but a place however lovely is only agreeable in solitude or good society neither of which charm does it possess. If you two marry Scotchmen, take care they are good ones. The Scotch are like foreigners in one respect, the very high *alone* are tolerable and they not always. . . .[70]

This harsh judgement, so different from the affectionate public regard Mrs. Somerville showed for Scotland, was due in large part to the peculiar circumstances of 1837, when to outward appearances the Somervilles seemed to be basking in good fortune — a secure post for William Somerville, a civil pension and fame for his wife, four grown children, their sons making names for themselves, a home and an exciting London life — and the private reality that added to this picture the loss of their small capital, a heavy debt, straitened circumstances, two undowered daughters, their failure to find patronage for young Greig, and their own advancing years.

Back in London she was immediately able to perform a useful service to William Whewell and to British science. Whewell and Lubbock had for some years been investigating tidal phenomena and from 1834 had received modest grants from the British Association toward their work in reducing tidal observations and establishing methods of analysis and prediction.[71] Whewell, picturing a series of observations made by mathematicians and scientists along a great circle of coastline, wrote colleagues in coast countries urging them to join in an effort so important to maritime nations.[72] While Mary Somerville was still in Scotland, he wrote to her at Chelsea, enclosing a memorandum on 'Tide observations to which subject I am desirous of drawing the attention of the Russian government'. He hoped that she, through her 'Russian friends' could bring the matter to the notice 'of the Administration of the Navy, so as to lead to some steps being taken, in the way of directing observations to be made'.[73] In all likelihood Whewell knew of her usefulness in the matter of R. W. Fox's magnetic instruments; Fox had written her on 30 April to report that the Russian government had ordered two of the devices, to be made in London, after her brother-in-law — to whom she had earlier written on the matter —called them to the attention of the Russian authorities.[74]

She was equally successful on Whewell's behalf. She sent Whewell's memorandum off to Greig, who would, she assured him, 'take a lively interest . . . in the Theory of the Tides which is so intimately connected with his profession'.[75] By August she was able to report, by letter to Whewell 'at the British Association/Newcastle', that the Admiral had forwarded the philosopher's request to the Imperial Academy of Sciences along with his own instructions to carry it out and that regular observations would now be made at all fixed establishments along the Russian coasts.[76] As the study of light and electricity had dominated physical science in the 1820s and 1830s, researches on terrestrial magnetism would be a major effort in the coming decades.

Whewell, in his letter of thanks to Mary Somerville for her intercession, declared:

... You would I think have enjoyed the "Assn week", as the Newcastle people called it; if your plans and engagements had led you into it; for everything (at least of a public nature) [— Babbage and Whewell had exchanged sharp words at a Committee meeting[77] —] went on very satisfactorily, and the arrangements were better than they have been before. Among the tasks we have proposed for the future, the most interesting perhaps is, a grand scheme of magnetic observations combined with an antarctic expedition, which we are to recommend to government. There seems the strongest probability that we shall be able, not only to determine the magnetic poles at present as a basis for future observations of their changes but also find a very easy method of determining the longitude, by means of the simultaneous changes of the variation. We are also to have a Reform Bill for the stars, improving the representation of the skies, putting a number of constellations in schedule A, and arranging the boundaries of others according to their star-populations. So you see we charge ourselves with earth and heaven at the same time.[78]

Mary Somerville would have enjoyed thoroughly being at hand to witness 'the Magnetic Crusade'[79] in which so many of her friends — Whewell, Herschel, Lubbock, Forbes, Wheatstone, Beaufort, James C. Ross (1800–1862; F.R.S., 1828), Thomas Spring-Rice (1790–1866; F.R.S., 1841) and, above all, Edward Sabine — took part over the next decade. But in September the Somervilles were to leave England; they did not realize at the time it was to be a permanent removal.

In July William Somerville had a severe attack of jaundice and for a fortnight was very ill. A London specialist was called in, declared the doctor must leave the unhealthy Chelsea house if he were to recover, and expressed surprise the whole family was not afflicted. Dr. and Mrs. Somerville went to James Somerville's small house in Savile Row, the girls stayed at Chelsea. All through August Somerville was very feeble and his doctor advised a winter in Rome. Mary Somerville, acting with her usual good sense in time of crisis, laid out their route, solicited letters of introduction from friends,[80] and packed their belongings at Chelsea. The bustle was considerable, and papers and possessions higgledy-piggledy. On 6 September the Board at Chelsea met, granted Dr. Somerville a six-month leave, and appointed Dr. James C. Somerville to act in his place.[81] On the morning of 19 September 1838 they sailed, expecting to return in the late spring of the following year.

CHAPTER 11

OUTSIDE THE MAINSTREAM OF SCIENCE

1. Italy, 1838–40

Rome was expensive and crowded in the winter of 1838. Mary Somerville wrote her son that 'according to the police report there are no less than eight thousand English besides a vast number of other foreigners'[1] in the city. Dr. Somerville 'on the whole' appeared to be improving. His wife, fatigued by her exertions, was less well for a time.

In London during the spring of 1838 Mrs. Somerville had sometimes appeared rather tired. Benjamin Disraeli (1804–1881; F.R.S., 1876), dining with a few scientific folk at Lord and Lady Powerscourt's on 15 March, thought 'the party . . . good, in some instances rather funny . . . Murchison a stiff geological prig, and his wife silent, Mrs. Somerville grown very old and not very easy . . .'.[2] A month later, however, she made a better impression on one of her Chelsea callers, Professor George Ticknor of Harvard, who had met her three years earlier. Ticknor, a detailed diarist, does not remark her health in his journal entry on this occasion but does declare her to be 'certainly among the most extraordinary women that have ever lived, both by the simplicity of her character and the singular variety, power and brilliancy of her talents'.[3]

Her spirits were brightened[4] — as were all in the English scientific community — by the return of Sir John Herschel and his family from South Africa in the spring of 1838. Sir John immediately resumed his busy scientific life, was honoured on 15 June with a great public dinner to mark his return, and took a leading part in the B.A.A.S. meeting at Newcastle in August. Both Dr. Somerville and Woronzow Greig were present at the public dinner — only gentlemen attended — and are listed among the distinguished guests.[5] Shortly afterward it was announced that Sir John had received one of the 'Coronation baronetcies' marking the crowning of the young Queen Victoria,[6] an attention paid him as a scientist and science as a sphere of activity. None of the open breaches between traditionalists and reformers in the Scientific Establishment that had characterized the late 1820s and early 1830s remained. The direction of science had been firmly fixed, as the reformers wished, toward increasing professionalism and Herschel would dominate British science in the next three decades.

He would also again assume the role of chief scientific adviser to Mary Somerville. During the Herschels' stay at the Cape of Good Hope, letters, gifts and books had continued to pass between the two households,[7, 8] but the sometimes day-to-day consultation that had characterized the association of Herschel and Mrs. Somerville in the late 1820s and early 1830s was no longer possible. Nor was there time or opportunity for them to have much scientific conversation before the Somervilles left for Italy. Their friendship and respect for each other, however, remained unimpaired, and as the years passed, she became more and more dependent on him for scientific news, advice and encouragement.

In the period between her visit to Scotland in 1837 and her departure for Italy in 1838, Mary Somerville did little science. She declined an invitation from Henry Bowditch[9] to write an elaborate review of his father's work on Laplace — Nathaniel Bowditch had died in Boston, Massachusetts on 16 March 1838 — fearing, as she put it, that she 'should not do justice to the memory of so great a man'.[10] For more than a decade she had been hard at authorship and the stresses of the past three years had added to her weariness.

In Rome, however, she had opportunity to rest. In early December she wrote her son that they had met

. . . with all the best people of every nation which is very amusing. The girls make a great figure here with their command of french italian and german and are . . . favourites. I think the life here very agreeable. I *generally write till two and then go to see* something, after that we make a few calls . . . walk on the Pincio where we meet most of our acquaintance . . . dine . . . and either go out in the evening or receive at home. . . .[11]

This basic pattern would obtain for the rest of their lives. The Somervilles' daily society would also change. Rather than moving in scientific circles — largely non-existent in Italy — their companions, as their letters indicate, were members of the Italian nobility, visiting English and other foreigners. Mary Somerville's scientific intercourse would be limited to a few of these and to letters.

At Christmas time Dr. Somerville had another attack of jaundice.[12] The rainy weather was thought to have brought it on but he was not well enough to travel south until April.[13] Naples, however, was soon too warm for them, and the family started northward on what they expected to be the first leg of a leisurely return journey home.

At Florence,[14] where they spent a fortnight, they were cordially received by their old friends, Professor G. B. Amici and his wife, both of whom recalled with gratitude the Somervilles' kindnesses to them during their visit to London over a decade earlier. Amici had come to Florence in 1831, at the invitation of the Grand Duke of Tuscany, to be inspector-general of education and head of the astronomical observatory and of the Royal Museum of Physics and Natural History. Tuscany in these years boasted the best scientific society to be found on the Italian peninsula. Its head of state, the cultivated Leopold II (1797–1870; F.R.S., 1838), was one of the most enlightened rulers in Italy and a generous patron of science. His library at the Pitti Palace had an excellent collection of scientific books and many current scientific periodicals. In addition to maintaining an observatory and museums, he undertook in

1838 the establishment of a Tuscan scientific society.[15] Leopold's interest in science was sincere and informed.

Amici and his colleague, the physicist Vincente Antinori (1792–1866), showed the Somervilles over the Pitti Palace galleries and its museums and made sure they saw everything 'that was interesting in science and art',[16] including instruments and telescopes once used by Galileo. Mrs. Somerville wrote her son that the Grand Duke, on hearing that they were in Florence

> . . . immediately sent a message to say that he hoped we would come to see him. Dr. S. and I accordingly went and spent nearly two hours with him in the most agreeable and easy conversation the more so as he was quite alone, there being nobody in the room but Mr. Antinori who went to present us. I have met with few people more generally well informed or more agreeable, his knowledge of Science for a Prince is very remarkable and his anxiety for the improvement of his country very great. He has given orders that I was to be invited to attend their first Scientific association which is to take place at Pisa, and just before I came away I received a very pretty letter[17] from him with a work just published on his improvements in the Maremma. . . .[18]

Leopold offered Mrs. Somerville permission to borrow books from his private library — a privilege hitherto enjoyed only by 'the four directors' and of great importance to her when she began in earnest to collect material for what she described as her 'long neglected Physical Geography still in embrio'.[19]

William Somerville's leave from Chelsea had been extended for another six months, so the family spent part of June and all of July at a small village near Lake Como. In August they moved to Baden to take the water, but the damp, cold climate of the region, together with upsetting letters about the Wemyss affair and the annual debt payment, so disturbed Dr. Somerville that he became seriously ill again. His family, convinced that warmth and peace of mind were necessary to his health, determined to return to Italy as soon as possible. The Board at Chelsea 'most handsomely and readily granted'[20] him an additional six-months leave.

The Somervilles' first plan was to winter at Genoa, which they reached in November, but they soon decided that Florence was preferable. William Somerville enlisted Amici's help in finding suitable lodgings.[21, 22] Even so, the doctor's recovery from jaundice was slow and their lives necessarily quiet. From Florence Mary Somerville wrote her son a week after her fifty-ninth birthday:

> . . . I am now in my 60th year!!! but thank God I am in better health and hope to be soon quite well . . . I think my eyes are better . . . I have got gout in both hands, my finger joints are swolled [sic] and I often find it difficult to hold my pen besides I have become *horribly* deaf so much so that I put all at home of patience and make strangers think me an idiot . . . It is too bad to be at once blind, deaf, and lame — you see what a useless old hag your mammy has become. . . .[23]

Her mind, however, remained vigorous and she continued her scientific writings, making good use of the Royal library. Surviving requests to Amici[24] document her interest in works on travel and geography[25] as she continued work on physical geography. To her friend Mrs. Leonard Horner she wrote that she had

. . . been busy taking advantage of the Grand Duke's permission of having books home from his library, which has never been granted [except to its directors] but once before. It is an excellent one and contains all modern publications, and transactions and they even offered to get [for me] any thing they had not. . . .[26]

In London in February 1840 John Murray published a fifth edition of *On the Connexion of the Physical Sciences.*[27] He printed only 750 copies.[28] It was a replica of the fourth (1838) edition, and this printing merely kept the book in stock. The fourth edition had brought Mary Somerville badly needed royalties of only £57.10.8 in 1839[29] and the fifth edition did no better. In its first year the return was £56.19.5[30] The high excitement that new editions had previously aroused was missing.

In the spring of 1840 William Somerville resigned his post at Chelsea. He did so with reluctance. He had been on leave at full pay. Now a new Governor and a new Board were anxious to make permanent arrangements for a physician. Somerville was put on half-pay at £600 a year. This act did not go unnoticed. A *Times* leader of 13 April 1840 began:

There is a Dr. Somerville whose wife is a very accomplished and very learned lady, and who has written on scientific subjects in a style that would not disgrace a professor at any university in Europe. This lady most deservedly got a pension of £200 a year. Her husband, anxious no doubt, to acquire some equal advantage has been this long time past a close and assiduous hanger-on among the Melbourne Ministers. . . .

The report went on to give a highly coloured account of his retirement procedure and ended, 'The whole transaction smells hugely of a foul Whig job'.[31] There was one letter to the *Times*[32] on the subject, then the sensation died down.

To escape the summer heat the family removed to Siena, reputedly cooler than Florence because of its greater altitude. To her friend Mrs. Horner Mary Somerville confessed:

. . . Florence is charming in everything but society, the English with a few exceptions have been very vulgar & disagreeable, so much so that we kept with our set, and the Italians are not fond of making acquaintances for a few months, besides very few of them receive at all. Then the gossip and scandal that go on are quite intollerable [*sic*], no one escapes and the worst constructions are put on the most simple actions especially among the English who seem to have too much to do with their neighbours affairs to mind their own. Fortunately we were independent, having some highly valued friends with whom we lived in daily intercourse . . . you may believe I regret going to Siena . . . [on account of leaving the Grand Duke's library], for though it [Siena] is a university I fear I shall not have the advantage of modern books.[33]

Away from Florence, lack of scientific reading matter and of good scientific society was an acute problem during all of Mary Somerville's Italian years. Amici, however, was a ready supplier of scientific materials and instruments during their stay in Siena.[34]

As the months passed, correspondence with English scientific friends grew more sporadic. Mrs. Somerville still provided letters of introduction to Italian philosophers visiting London, but her immediate knowledge of London science was far from complete. Only the first sheet of a letter to Charles Babbage introducing the Tuscan mathematician Corridi (1806–1878) survives, but it carries a poignant plea to make the visitor 'acquainted with the state of science in England',[35] an acquaintance Mary Somerville realized she no longer had. She hoped that some of her friends would be in Turin in early autumn at the second great scientific congress sponsored by the new Tuscany association and that they would find their way to Siena for a visit with her. She assured Mrs. Horner that Charles and Mary Lyell '. . . could not make a more agreeable trip than to Turin . . .'.[36]

On learning that Babbage would be at the congress, she wrote him:

I take it for granted . . . that you will come to Siena as soon as the meeting at Turin is over, for surely you could not be so near without coming to see us . . . I have a thousand things to ask you and to learn — After being so long out of the world you will find us in a deplorable state of idleness and ignorance for we have completely adopted the dolce far niente of this country; life is a pleasure in this heavenly climate without seeking anything more. . . .[37]

Babbage planned a visit, but while at Turin became ill and had to return directly to England, a great disappointment to all.[38]

As winter approached, Dr. Somerville began to feel the cold more acutely. The family decided to exchange the scientific advantages of Florence for the warmer climate of Rome.[39] For the next few years the Somervilles would make that city their principal residence.

2. And After . . .

The wonder is that Mary Somerville did any science at all after leaving England. But she continued with her studies and her work, under great disadvantage and far removed from the scene of discoveries and researches. She brought out four more editions of *On the Connexion of the Physical Sciences* in her lifetime — in 1842, 1846, 1849 and 1858 — and a tenth appeared in 1877, five years after her death. Some fifteen thousand copies of the book were printed and sold over four decades.

In 1845 she carried out her third piece of experimental work, a study of the effect of sunlight on vegetable juices. An extract of her letter to Herschel about these experiments was published in the *Abstracts of the Philosophical Transactions* in that year.[40]

In 1848 Mrs. Somerville published her third book, *Physical Geography*,[41] in two volumes. A pioneering work, it anticipated a regional approach to geography rather than one made solely along national or political lines. Five more editions appeared in her lifetime (1849, 1851, 1858, 1862, 1870) and a seventh in 1877. In all some sixteen thousand copies were sold.

She began her fourth work, *On Molecular and Microscopic Science*,[42] in her eighty-fifth year. It was published in 1869, in her eighty-ninth. Its first volume considered small bodies of inanimate matter (molecules), its second small bodies of animate matter (microscopic organisms). Fifteen hundred copies were printed. Reviewers tended to dwell on the age and distinction of its author rather than on its contents, but the work did in fact touch on some important new propositions. It avoided, however, any mention of Darwin's evolutionary ideas — Mary Somerville believed they needed further experimental investigation — even though Darwin supplied some of the plates for illustrations.[43]

Her last work was an autobiography, begun in her eighty-ninth year and completed before her death in 1872. A heavily edited version of this work was published in 1873, a fourth thousand brought out in 1874. Her daughter Martha, advised by their friend Frances Power Cobbe (1822–1904), omitted from the printed life markedly scientific sections of the manuscript, as well as many references to persons and events they judged uninteresting or unsuitable as contents. A selection of letters to Mary Somerville from eminent persons was woven into the text. No critical examination of the account was undertaken; its chronology is wavering and events occasionally confused. *Personal Recollections* emerges with many Victorian touches that would have been foreign to Mary Somerville herself.

Except for two visits to England — in 1844 and again in 1848 — and a year in Germany (1849), Mary Somerville remained for the rest of her life in Italy. She had no settled home — the family wandered about the peninsula, living for weeks or months or years in Florence, Rome, Bologna, Genoa, Turin, Spezia and Naples. William Somerville died in 1860, in his nintieth year. The greatest sadness of Mary Somerville's eventful life was the loss of her son in 1865. Neither of the two Somerville daughters married. They survived their mother but a few years. Her heir was her soldier nephew, Sir William George Herbert Taylor Fairfax, second baronet (1831–1902), the son of her soldier brother Henry.

Eleven Italian scientific societies and two American societies made Mrs. Somerville an honorary member in the years 1840 to 1870. In 1869 she received the Patron's Medal of the Royal Geographical Society. Her old friend Sir Roderick I. Murchison was President of the Society at the time. She outlived all the scientific friends of her London years except Sir G. B. Airy, R. W. Fox, Sir Henry Holland, Sir Charles Lyell, J. B. Pentland, Sir John Rennie, Sir Edward Sabine, Adam Sedgwick and Sir Charles Wheatstone.

Mary Somerville died peacefully on 29 November 1872, in her ninety-second year. The day before her death she amused herself working on a paper on quarternions. She is buried in the English Cemetery in Naples.

A GUIDE TO NOTES AND CITATIONS

All items in the Somerville Collection deposited in the Bodleian Library are arranged by a shelf number and a filing code as follows: 'Dep' plus 'b.205' or 'c.NNN', where NNN = 351–378. Since the designation 'Dep.' is common to all items, it is omitted in the citations, which are thus made in this form: 'c.359,17,MSIF-3', followed by the identification of the item and its date, if any. The 'c.359' represents the shelf-mark. The '17' is a box number. 'MSIF' is a code identifying the nature of the contents of the box, e.g., inner family papers. The last number, '3', identifies the folder in which the item is located. Other abbreviations used are:

MS — Mary Somerville
WS — William Somerville
TS — The Reverend Dr. Thomas Somerville
WG — Woronzow Greig
Martha S — Martha Charters Somerville
Mary CS — Mary Charlotte Somerville
OCPS — *On the Connexion of the Physical Sciences* (various editions, each identified by date)
PR — *Personal Recollections, from Early Life to Old Age, of Mary Somerville, with Selections from her Correspondence* (London, 1873)
Lit. Gaz. — *Literary Gazette* (London)
BE — Biblioteca Estense (Modena, Italy)
BL — British Library (London)
NLS — National Library of Scotland (Edinburgh)
RS — Library of the Royal Society of London

Names in brackets (e.g., [David Brewster]) are based on internal evidence or reputable attribution.

Dates in brackets (e.g., [1832]) are based on internal evidence.

NOTES

Chapter 1 — Scottish Beginnings
pages 1–10

1. *The Morning Post* (London), 2 Dec. 1872.
2. Martha Somerville (ed.), *Personal Recollections from Early Life to Old Age, of Mary Somerville* (London, 1873), p. 25.
3. *ibid.*, p. 61.
4. c.355,5,MSAU-2: pp. 29–30.
5. PR, pp. 27–28, 88.
6. *ibid.*, p. 37.
7. c.364,14,MSIF-26: Mrs. Janet Somerville Pringle Elliot to WG, 11 Feb. 1837.
8. PR, pp. 75–76.
9. *ibid.*, p. 80.
10. Either James Ferguson's *Astronomy Explained upon Sir Isaac Newton's Principles, and made Easy for those who have not studied Mathematics* (London, 1756) or his *The Young Gentleman's and Lady's Astronomy* (London, 1768).
11. PR, p. 78.
12. c.355,5,MSAU-2: p. 57.
13. *ibid.*
14. *ibid.*
15. *ibid.*
16. c.373,22,MSW-1: William Wallace to MS, 12 May 1816; 18 May 1816.
17. PR, p. 82.
18. c.375,25,MSDIP-1: John Wallace to MS, 12 July 1811.
19. PR, p. 72.
20. c.360,10,MSFP-47: TS's MS. 'Almanac'.
21. NLS 11092 (Minto Papers).f.129: WS to Sir Gilbert Elliot, 4 Feb. 1791.
22. c.360,10,MSFP-47: TS's MS. 'Almanac'.
23. NLS 1151.f.131: WS to Lord Minto, 12 Mar. 1798.
24. c.360,10,MSFP-47: TS's MS. 'Almanac'.
25. c.377,27,MSWS-1: WS to Martha S, 4 Feb. 1848.
26. *Sketches Representing the Native Tribes, Animals, and Scenery of Southern Africa, from Drawings Made by the Late Samuel Daniell, Engraved by William Daniell* (London, 1820).
27. c.377,27,MSWS-2: WS's MS. account of 1800 expedition.
28. c.377,27,MSWS-3: WS's MS. account of 1801–1802 expedition.
29. John Barrow, *Voyage to Cochin China* (London, 1806), Appendix.
30. c.359,9,MSFP-38: Samuel C. Somerville to WS, 19 Feb. 1810.
31. W. Munk, *The Roll of the Royal College of Physicians of London* (2nd edn., London, 1878), iii. 306.
32. D. W. Krüger and C. J. Beyers (eds.), *Dictionary of South African Biography* (Johannesburg, 1977), iii. 748–749.
33. c.377,27,MSWS-1: Fragment, MS. autobiography of WS.
34. c.360,10,MSFP-47: TS's MS. 'Almanac'.
35. *ibid.*
36. c.355,5,MSAU-2: p. 59.
37. PR, p. 58.

38. c.370,20,MSH-4: Sir William Herschel to William Wallace, 8 July 1812.
39. Munk's *Roll*, iii. 79–80.
40. PR, pp. 79–80.
41. *ibid.*
42. c.355,5,MSAU-2: p. 61.
43. c.355,5,MSAU-2: pp. 64–65.
44. *Transactions of the Royal Society of Edinburgh,* vii (1815), 543.
45. *Scots Magazine,* lxxvi (May 1814), 398.
46. c.360,10,MSFP-47: TS's MS. 'Almanac'.
47. *ibid.*
48. NLS 9236 (John Thomson Papers).f.65: John Thomson to his wife Margaret, 25 July 1815.
49. NLS 9236.f.72: John Thomson to his wife Margaret, 31 July 1815.
50. NLS 9236.f.74: John Thomson to his wife Margaret, 20 Aug. 1815.
51. *Army List* (January 1816), p. 99.

Chapter 2 — London Beginnings
pages 11–17

1.   c.360,10,MSFP-44: MS to TS, 24 June 1816.
2.   Royal Institution Archives: Managers' Minutes, 5 Feb. 1816.
3.   NLS 9818 (Leonard Horner Papers).f.60: Leonard Horner to A. J. G. Marcet, 14 Mar. 1816.
4.   K. M. Lyell, *Memoir of Leonard Horner, F.R.S., F.G.S., Consisting of Letters to his Family and from Some of his Friends* (London, 1890), i. 92–93.
5.   A. Garrod, 'Alexander John Gaspard Marcet', *Guy's Hospital Reports* (Oct. 1925), 373–387.
6.   A. J. G. Marcet, *Essay on the Chemical and Medical Treatment of Calculus Disorders* (London, 1817).
7.   L. P. William, *Michael Faraday: A Biography* (New York, 1965), p. 19.
8.   [Jane Marcet], *Conversations on Chemistry* (London, 1806).
9.   E. F. Smith, *Old Chemistries* (New York, 1927), p. 68.
10. L. P. Williams (ed.), *The Selected Correspondence of Michael Faraday* (Cambridge, 1971), ii. 914.
11. Harriet Martineau, *Biographical Sketches* (2nd edn., London, 1869), p. 387.
12. Jasper Ridley, *Lord Palmerston* (London, 1970), p. 85.
13. NLS 9818.f.62: Leonard Horner to A. J. G. Marcet, 24 Apr. 1816.
14. c.360,10,MSFP-44: MS to TS, 24 June 1816.
15. c.371,21,MSJ-1: Robert Jameson to WS, 18 May 1816.
16. c.360,10,MSFP-44: MS to TS, 1 Apr. 1817.
17. *ibid.*
18. c.360,10,MSFP-44: MS to TS, 12 Dec. 1816.
19. H. B. Woodward, *The History of the Geological Society of London* (London, 1907), p. 278.
20. *List of the Linnean Society of London* (1817), p. 9.
21. RS Certificate Ledger, 1810–1820: William Somerville. Elected 11 Dec. 1817.
22. c.371,21,MSJ-1: Robert Jameson to MS, 25 Apr. 1816.
23. Robert Jameson, *System of Mineralogy* (2nd edn., Edinburgh, 1816).
24. c.371,21,MSJ-1: Robert Jameson to MS, 18 Mar. 1817.
25. c.372,22,MSW-1: William Wallace to MS, 12 May 1816; 16 May 1816.
26. PR, pp. 105–106.
27. c.356,6,MSAU-4: MS. notes for autobiography.
28. *ibid.*
29. c.369,19,MSB-8: J. B. Biot to WS, 1 June 1817.
30. c.369,19,MSB-8: J. B. Biot to MS, 27 June 1817.
31. c.371,21,MSJ-1: Robert Jameson to MS, 18 Mar. 1817.

NOTES

201

Chapter 3 — The First Trip Abroad
pages 19–33

1. c.355,5,MSAU-1: MS's MS. diary of her first Continental tour, 1817–1818.
2. Harold Hartley, *Humphry Davy* (London, 1966), pp. 100–108.
3. Maurice Crosland, *The Society of Arcueil* (London, 1967), pp. 272–276, 410–411, 434,
4. E. C. Patterson, *John Dalton and the Atomic Theory* (New York, 1970), pp. 204–212.
5. c.357,7,MSFP-4: MS to Lady Fairfax, 18 Aug. 1817.
6. A. Lacroix, 'Une Familie de Bons Serviteurs de l'Académie des Sciences et du Jardin des Plantes', *Bulletin du Muséum,* 2⁰ Serie, x (1938).
7. c.355,5,MSAU-1: 23 July 1817.
8. Crosland, *Society of Arcueil,* p. 126.
9. *ibid.*, pp. 95–96.
10. J. B. Biot, *Physique Mécanique par E. G. Fischer . . . traduite de l'allemand . . . par M. Biot* (Paris, 1806).
11. c.355,5,MSAU-1: 26 July 1817.
12. c.355,5,MSAU-1: 24 July 1817.
13. c.355,5,MSAU-1: 25 July 1817.
14. *ibid.*
15. c.355,5,MSAU-1: 26 July 1817.
16. *ibid.*
17. *ibid.*
18. *ibid.*
19. *ibid.*
20. Lady Morgan, *France* (London, 1817).
21. c.355,5,MSAU-1: 26 July 1817.
22. c.355,5,MSAU-1: 27 July 1817.
23. J. A. H. Lucas, *Méthodique des Espèces Minérales,* Part I (Paris, 1806); Part II (Paris, 1812).
24. c.355,5,MSAU-1: 30 July 1817.
25. *ibid.*
26. c.355,5,MSAU-1: 31 July 1817.
27. *ibid.*
28. c.355,5,MSAU-1: Loose sheets.
29. c.355,5,MSAU-1: 1 Aug. 1817.
30. *ibid.*
31. c.355,5,MSAU-1: 2 Aug. 1817.
32. Crosland, *Society of Arcueil,* p. 335.
33. Hudson Gurney, *Memoir of Thomas Young, M.D., F.R.S.* (London, 1831), p. 32.
34. c.355,5,MSAU-1: 2 Aug. 1817.
35. c.355,5,MSAU-1: 5 Aug. 1817.
36. PR, pp. 108–113.
37. *ibid.*, p. 108.
38. *ibid.*
39. *ibid.*, pp. 108–109.
40. *ibid.*, p. 113.
41. c.357,7,MSFP-4: MS to Mrs. Janet Somerville Pringle, 19 Oct. 1817.
42. PR, p. 114.
43. c.357,7,MSFP-4: MS to Lady Fairfax, 18 Aug. 1817.
44. c.357,7,MSFP-4: MS to Mrs. Janet Somerville Pringle, 19 Oct. 1817.
45. PR, p. 117.
46. c.360.10,MSFP-44: MS to TS, 24 Mar. 1818.
47. c.357,7,MSFP-4: MS to Lady Fairfax, 12 Apr. 1818.
48. c.357,7,MSFP-4: MS to Mrs. Janet Somerville Pringle, 19 Oct. 1817.
49. PR, p. 117.
50. Williams, *Faraday,* p. 38.

51. 'Scientific Intelligence from Italy', *Scots Magazine*, lxxix (Aug. 1817), 69.
52. c.371,21,MSM-5: D. P. Morichini to MS, 29 July 1826.
53. c.355,5,MSAU-2: p. 86.
54. Woodward, *History of the Geological Society*, p. 280.
55. c.355,5,MSAU-2: p. 86.
56. BE, Amici Correspondence, 7188: WS to G. B. Amici, 9 July 1819.
57. PR, pp. 113–114.
58. Lyell, *Memoir of Leonard Horner*, i. 148–150.
59. PR, p. 114.
60. *ibid.*, p. 118.
61. *ibid.*, p. 126.
62. *ibid.*, pp. 118–119.
63. c.355,5,MSAU-2: pp. 91–92.
64. c.357,7,MSFP-4: MS to Mrs. Janet Somerville Pringle, 26 Dec. 1817.
65. c.357,7,MSFP-4: MS to Lady Fairfax, 8 Jan. 1818.
66. PR, pp. 125–126.
67. Verney Family Papers (Claydon House, Bucks.): Benjamin Smith to his sister Fanny Smith, 27 Mar. 1818.
68. c.355,5,MSAU-2: p. 102.
69. PR, p. 124.
70. c.357,7,MSFP-4: MS to Lady Fairfax, 12 Apr. 1818.
71. c.355,5,MSAU-1: p. 96.
72. c.370,20,MSG-1: Josephine Gay-Lussac to MS, 23 Feb. 1819.
73. c.370,20,MSD-2: A. P. de Candolle to MS, 5 June 1819.
74. BE 7188: WS to G. B. Amici, 9 July 1819.
75. BE 7190: G. B. Amici to WS, 1 Aug. 1819.
76. c.360,10,MSFP-46: TS to WS, 17 July 1819.
77. D. Peterkin and M. Johnstone, *Commissioned Officers in the Medical Services of the British Army 1660–1960* (London, 1968), i. 96.
78. c.360,10,MSFP-43: TS to MS, n.d. [June or July 1819].
79. c.360,10, MSFP-44: MS to TS, 28 Sept. 1819.
80. *ibid.*
81. Peterkin and Johnstone, *Commissioned Officers*, i. 96.
82. c.360,10, MSFP-44: MS to TS, 28 Sept. 1819.
83. c.370,20,MSC-5: Clot Bey to WS, 2 Jan. 1834.
84. c.375,25,MSDIP-6: Académie Royale de Médecine (Paris) to WS, 31 Mar. 1834.
85. B. B. Cooper, *Life of Sir Astley Paston Cooper* (London, 1843), ii. 239–248.
86. c.373,23,MSBUS-8: State of Accounts, 5 June 1843.

Chapter 4 — In the Mainstream of London Science
pages 35–53

1.   c.355,5,MSAU-3: p. 147.
2.   George Jones, *Sir Francis Chantrey, R.A.: Recollections of his Life, Practice & Opinions* (London, 1849), p. 98.
3.   PR, p. 106.
4.   Archibald Geikie, *Annals of the Royal Society Club* (London, 1917), pp. 268–302.
5.   *Lit. Gaz.*, 546 (1827), 730.
6.   PR, p. 130.
7.   John Brinkley, *Elements of Astronomy* (Dublin, 1813).
8.   c.371,21,MSK-1: Henry Kater to MS, 17 Nov. 1821.
9.   PR, pp. 128–129.
10. c.372,22,MSW-4: W. H. Wollaston to MS, 19 Mar. n.y. [1824].
11. PR, p. 148.
12. *ibid.*

13. c.372,22,MSW-4: W. H. Wollaston to MS, n.d. ('It occurs to me that in trying to observe a star . . .').
14. c.372,22,MSW-4: W. H. Wollaston to MS, n.d. ('Friday morng. 5th').
15. PR, pp. 133–134.
16. Woodward, *History of the Geological Society*, p. 306.
17. Geikie, *Annals of the Royal Society Club*, p. 274.
18. *ibid.*, p. 284.
19. c.360,10, MSFP-44: MS to TS, 22 Dec. 1820.
20. L. F. Gilbert, 'The Election to the Presidency of the Royal Society in 1820', *Notes and Records of the Royal Society of London*, ii (1954–55), 259–262.
21. c.370,20,MSD-2: Humphry Davy to WS, 4 Dec. n.y. [early 1820s].
22. Unpublished letter owned by Mrs. C. Colvin of Oxford: Harriet Edgeworth to Lucy Edgeworth, 11 Apr. 1822.
23. C. Colvin, *Maria Edgeworth — Letters from England 1813–1844* (Oxford, 1971), pp. 320–325.
24. The writer Anna Letitia Barbauld (1743–1825).
25. Henrietta (Harriet) Beaufort (1778–1865), sister of the hydrographer Admiral Sir Francis Beaufort.
26. The Scottish poetess and dramatist Joanna Baillie (1762–1851) and her younger sister Agnes. Their brother Dr. Matthew Baillie (1761–1823; F.R.S., 1790), the anatomist, was also a friend of the Somervilles.
27. Colvin, *Edgeworth Letters from England*, pp. 321–322.
28. Henrietta (Harriet) Edgeworth (1801–1889), who in 1826 married the Revd. Richard Butler of Trim.
29. Smith refers to Mrs. Somerville's *Preliminary Dissertation* (1832).
30. Colvin, *Edgeworth Letters from England*, p. 601.
31. *ibid.*, pp. 383–384.
32. Unpublished letter owned by Mrs. C. Colvin of Oxford: Harriet Edgeworth to Mrs. Edgeworth, 26 May 1822.
33. NLS 9236,f.117: Isabella Thomson to William Thomson, 18 Aug. 1822.
34. Lyell, *Memoir of Leonard Horner*, i. 203–204.
35. c.372,22,MSW-1: James Veitch of Inchbonny to MS, 12 Oct. 1836.
36. J. G. Lockhart, *Narrative of the Life of Sir Walter Scott, Bart.* (London, 1848), ii. 57.
37. c.370,20,MSD-2: Letter in French to MS, 29 May n.y. [early 1820s], signature illegible.
38. c.360,10, MSFP-44: MS to TS, 22 Dec. 1820.
39. PR, p. 106.
40. c.355,5,MSAU-3: pp. 76–77.
41. RS Certificate Ledger, 1821–1830: Isambard Kingdom Brunel. Elected 10 June 1830.
42. c.355,5,MSAU-3: p. 77.
43. c.355,5,MSAU-2: pp. 68–75.
44. c.355,5,MSAU-3: pp. 72–82.
45. c.355,5,MSAU-3: p. 73.
46. c.355,5,MSAU-3: pp. 80–81.
47. c.355,5,MSAU-3: p. 73.
48. c.355,5,MSAU-3: p. 78.
49. c.355,5,MSAU-3: p. 65.
50. PR, p. 104.
51. c.355,5,MSAU-2: p. 71.
52. c.355,5,MSAU-3: p. 74.
53. PR, pp. 135–139.
54. c.360,10,MSFP-44: MS to TS, 22 Dec. 1820.
55. *ibid.*
56. PR, pp. 136–137.
57. *ibid.*, p. 89.
58. c.371,21,MSP-1: W. E. Parry to MS, n.d. [Apr. or May 1824].

59. PR, p. 137.

60. c.372,22,MSS-1: Edward Sabine to MS, 8 Apr. n.y. [1832].

61. c.370,20,MSF-2: John Franklin to MS, 14 May n.y. [1826].

62. c.369,19,MSB-2: George Back to WS, 2 Feb. 1838; 10 Feb. 1838; 12 Feb. 1838.

63. *List of Carthusians 1801 to 1879* (London, 1879), i. 102.

64. J. A. Venn, *Alumni Cantabrigensis, Part II. 1751–1900* (Cambridge, 1942), iii.

65. Munk's *Roll*, iii. 306.

66. c.360,10,MSFP-44: MS to TS, n.d. ('Sunday') [late summer 1821 or 1822].

67. *ibid.*

68. *Gentleman's Magazine*, xciii, Part I (1823), 651.

69. c.360,10,MSFP-44: MS to TS, 10 Oct. 1823.

70. c.360,10,MSFP-46: TS to WS, 25 Aug. 1823.

71. c.357,7,MSFP-2: Lady Fairfax to WS, 20 Sept. 1823.

72. c.355,5,MSAU-3: p. 123.

73. PR, p. 157.

74. *ibid.*, p. 156.

75. c.371,23,MSM-4: Gerard Moll to MS, 10 Feb. 1826.

76. PR, p. 159.

77. c.370,20,MSH-1: Basil Hall to MS, 10 Oct. 1823; 9 Apr. 1824.

78. c.371,21,MSM-1: Sir James Mackintosh to WS, 29 Jan. 1829.

79. PR, pp. 141–151, 219–223.

80. W. Heberden, 'An Account of the Heat of July, 1825; together with some Remarks upon sensible Cold', *Abstracts of the Papers Printed in the Philosophical Transactions of the Royal Society of London*, ii (1815 to 1830), 260.

81. Williams, *Faraday*, p. 38.

82. [J. F. W. Herschel], 'Mrs. Somerville's Mechanism of the Heavens', *Quarterly Review*, xc (Sept. 1832), 548.

83. c.372,22,MSW-4: W. H. Wollaston to MS, n.d.

84. 'X. On the magnetizing power of the more refrangible solar rays. By Mrs. M. Somerville. Communicated by W. Somerville, M.D., F.R.S. Feb. 2, 1826', *Phil. Trans.* (1826), Part i, 132–139.

85. M. C. Harding (ed.), *Correspondance de H. C. Oersted avec divers savants* (Copenhagen, 1920), ii. 423–424.

86. Geikie, *Annals of the Royal Society Club*, pp. 278–280.

87. Harding, *Oersted Correspondance*, i. 289–290.

88. c.370,20,MSG-1: J. L. Gay-Lussac to WS, 14 Apr. 1826.

89. c.371,21,MSL-1: P. S. Laplace to MS, 28 Apr. 1826.

90. *Bulletin des sciences mathématiques, astronomiques, physiques et chimiques*, v (1826), 169.

91. c.371,21,MSM-5: D. P. Morichini to MS, 20 June 1826.

92. *Annales de chimie et de physique*, xxxiii (1826), 333–335.

93. *Zeitschrift für Physik und Mathematik*, vi (5) [1826], 265.

94. *Bulletin des sciences . . . .*, viii, Sec. i (1827), 244–247.

95. P. Reiss and L. Moser, 'Ueber die magnetisirende Eigenschaft des Sonnenlichts', Poggendorff's *Annalen . . .* , xcii (1829), 563–592.

96. c.355,5,MSAU-2: p. 104.

97. RS Herschel Papers, 16.383: WS to J. F. W. Herschel, 17 July 1826.

98. RS Herschel Papers, 16.384: WS to J. F. W. Herschel, 2 Aug. 1826.

99. RS Herschel Papers, 16.385: J. F. W. Herschel to WS, 4 Aug. 1826.

100. c.355,5,MSAU-2: p. 104.

101. Mary Somerville, '"On the Action of the Rays of the Spectrum on Vegetable Juices:" being an Extract from a Letter by Mrs. M. Somerville to Sir John F. W. Herschel, Bart., dated Rome, September 20, 1845. Communicated by Sir John F. W. Herschel, Bart., F.R.S.', *Abstracts of the Papers Communicated to the Royal Society of London from 1843 to 1850*, v (1861), 569–570.

102. c.371,21,MSL-2: P. S. Laplace to MS, 19 Aug. 1824.

103. c.371,21,MSM-1: François Magendi to MS, n.d. [1827].

104. c.369,19,MSB-13: Henry Brougham to WS, 27 Mar. 1827.

105. c.355,5,MSAU-2: p. 57.

106. c.369,19,MSB-13: Henry Brougham to WS, 26 Dec. 1826.

107. J. N. Hays, 'Science and Brougham's Society', *Annals of Science*, xx (Sept. 1964), 227–241.

108. c.369,19,MSB-13: Henry Brougham to MS, n.d. ('Satur^y').

109. PR, pp. 162–163.

110. *ibid.*

111. *ibid.*, p. 163.

112. c.369,19,MSA-1: G. B. Amici to WS, 28 June 1828.

113. *ibid.*

114. Somerville College (Oxford), Somerville Memorabilia: MS to Mrs. Frederica Sebright Franks, 4 Aug. n.y. [1827].

115. c.360,10,MSFP-46: TS to WS, 14 May 1827.

116. Venn, iii.

117. c.364,14,MSIF-20: WG to MS, 20 June 1828.

118. Somerville College (Oxford), Somerville Memorabilia: MS to Mrs. Fredrica Sebright Franks, 4 Aug. n.y. [1827].

119. Letters to the Somervilles at this address from January through May 1828 survive. See c.372,22,MSW-1: Henry Warburton to MS, postmarked 14 Jan. 1828 and c.370,20,MSD-2: Lady Davy to MS, n.d. ('Thursday') [probably 15 or 22 May 1828].

120. C. C. F. Greville, *Memoirs 1814–1860*, eds. L. Strachey and R. Fulford (London, 1938), i. 15.

121. c.372,22,MSW-4: Statement of Wollaston's bequest, dictated by W. H. Wollaston and dated 7 Dec. 1828.

122. RS Herschel Papers. 16.351: J. F. W. Herschel to MS, 6 Mar. 1848.

123. W. Babington, B. C. Brodie and J. C. Somerville, 'Appearances observed on inspecting the Body of the late William Hyde Wollaston, M.D., F.R.S. [December 24, 1828]', *London Medical Gazette*, iii (1829), 293–294.

124. c.370,20,MSC-5: Adm. Sir Edward Codrington to WS, 20 Nov. 1828.

125. c.370,20,MSH-1: Basil Hall to MS, 12 Dec. 1828.

126. c.372,22,MSS-8: William Sotheby to WS, 24 Dec. 1828.

127. c.363,13,MSIF-17: MS to Agnes G. Greig (Mrs. Woronzow), n.d. ('Monday') Mar. 1870.

128. c.372,22,MSY-1: Thomas Young to MS, 30 Oct. 1828.

129. At Oxford Peel took a double first in classics and mathematics in 1808.

130. RS Herschel Papers, 16.383: WS to J. F. W. Herschel, 17 July 1826.

131. c.369,19,MSB-1: Charles Babbage to MS, 14 Jan. n.y. [1829].

132. c.355,5,MSAU-2: pp. 112–113.

133. PR, pp. 129–136.

134. William Cockburn, *A Letter to Professor Buckland* (London, 1838).

135. William Cockburn, *A Remonstrance, Addressed to His Grace the Duke of Northumberland, upon the Dangers of Peripatetic Philosophy* (London, 1838).

136. William Buckland, *The Creation of the World, Addressed to R. I. Murchison and Dedicated to the Geological Society* (London, 1840).

137. University of St. Andrews Library, James David Forbes Papers: David Brewster to J. D. Forbes, 11 Sept. 1829.

Chapter 5 — The Mechanism of the Heavens
pages 55–93

1.   Charles Babbage, *Reflections on the Decline of Science in England and on Some of its Causes* (London, 1830).

2.   Henry Lyons, *The Royal Society 1660–1940* (London, 1968), pp. 245–246.

3.   Francis Baily, *Remarks on the Present Defective State of the 'Nautical Almanac'* (London, 1829).

4.   James South, *Refutation of the Numerous Mis-Statements and Fallacies Contained in a Paper Presented to the Admiralty by Dr. Thomas Young, Etc.* (London, 1829).

5.   James South, *Reply to a Letter in the Morning Chronicle Relative to the Interest Which the British Government Evinces in the Promotion of Astronomical Science* (London, 1829).

6.   J. E. Dreyer, H. H. Turner *et al.*, *History of the Royal Astronomical Society 1820–1920* (London, 1923), pp. 58–60.

7.   Harding, *Oersted Correspondance*, ii. 282.

8.   Babbage, *Reflections on the Decline of Science*.

9.   *ibid.*, pp. vi–ix.

10.   *ibid.*, p. xiv.

11.   *The Times*, 8 May, 22 May, 28 May, 29 May, 31 May, 26 June, 8 July, 15 July 1830.

12.   *ibid.*, 8 May 1830.

13.   *Athenaeum*, 134 (1830), 305–306.

14.   *Lit. Gaz.*, 694 (1830), 309.

15.   [David Brewster], 'Reflections on the Decline of Science in England . . .', *Quarterly Review*, xliii (1830), 305–342.

16.   Philip Ziegler, *King William IV* (London, 1971), p. 151.

17.   *The Times*, 5 Aug. 1830.

18.   Mollie Gillen, *Royal Duke: Augustus Frederick, Duke of Sussex (1773–1843)* (London, 1976), p. 153.

19.   A. C. Todd, *Beyond the Blaze: A Biography of Davies Gilbert* (Truro, 1967), p. 239.

20.   Gillen, *Royal Duke*, p. 182.

21.   *ibid.*

22.   Todd, *Beyond the Blaze*, pp. 242–243.

23.   William Jerdan, 'Royal Patronage of Great Public and National Institutions', *Lit. Gaz.*, 708 (1830), 531–532.

24.   [W. H. Fitton], *A Statement of Circumstances Connected with the Late Election for the Presidency of the Royal Society* (London, 1831), p. 34.

25.   *ibid.*, p. 35.

26.   *ibid.*, p. 36.

27.   *ibid.*, p. 37.

28.   *Lit. Gaz.*, 723 (1830), 769.

29.   Sir James South, *On the Proceedings of the Royal Society, &c., the Necessity of a Reform of its Conduct, and a re-modelling of its Charter, &c.* (London, 1830).

30.   Sir James South, *Charges Against the President and Council of the Royal Society* (London, 1830).

31.   A. B. Granville, *Science Without a Head* (London, 1830).

32.   RS Herschel Papers, 2.153: J. F. W. Herschel to Charles Babbage, 22 May 1830.

33.   RS Herschel Papers, 2.152: J. F. W. Herschel to Davies Gilbert, 1 July 1830.

34.   *Lit. Gaz.*, 717 (1830), 674.

35.   [Fitton], *A Statement*, pp. 16–17.

36.   *ibid.*, pp. 34–44.

37.   The signers of the requisition were, in this order, H. T. Colebrook, W. H. Fitton, Francis Baily, J. F. W. Herschel, Charles Stokes, W. J. Broderip, J. E. Bicheno, William Prout, Nathaniel Wallich, James Franck, Francis Beaufort, Edward Troughton, George Dollond, B. C. Brodie, W. F. Chambers, John Lindley, W. H. Pepys, William Allen, Thomas Horsfield, Herbert Mayo, G. B. Greenough, Leonard Horner, R. I. Murchison, William Swainson, Thomas Bell, Charles Macintosh, W. J. Hooker, William Whewell, Henry Coddington, Richard Willis, Francis Chantrey, and William Somerville. [Fitton], *A Statement*, p. 17.

38.   *ibid.*

39.   *ibid.*, p. 18.

40.   Todd, *Beyond the Blaze*, p. 258.

41.   [Fitton], *A Statement*, p. 18.

42. Todd, *Beyond the Blaze*, p. 258.
43. *The Times*, 23 Nov. 1830.
44. Granville, *Science Without a Head*, pp. 104–105.
45. Lyons, *The Royal Society*, p. 250.
46. RS Herschel Papers, 2.257: J. F. W. Herschel to Charles Babbage, 26 Nov. 1830.
47. RS Herschel Papers, 2.258: J. F. W. Herschel to Charles Babbage, 27 Nov. 1830.
48. *The Times*, 25 Nov. 1830.
49. [Fitton], *A Statement*, p. 27.
50. *The Times*, 29 Nov. 1830.
51. *The Times*, 30 Nov. 1830.
52. '(Advertisement)/ROYAL SOCIETY./ The undersigned Fellows of the Royal Society, being of the opinion that Mr. Herschel, by his varied and profound Knowledge and high personal character, is eminently qualified to fill the office of President, and that his appointment to the Chair of the Society would be peculiarly acceptable to men of science in this and foreign countries, intend to put him to nomination at the election this day. November 30, 1830.'

The advertisement was signed by Charles Babbage, Francis Baily, Peter Barlow, Edward Barnard, Francis Beaufort, Charles Bell, Thomas Bell, John Bell, James Ebenezer Bicheno, David Brewster, Thomas M. Brisbane, William John Broderip, Benjamin Collins Brodie, Edward Ffrench Bromhead, Henry James Brooke, Robert Brown, Mark Isambard Brunel, Isambard Kingdom Brunel, Thomas Catton, William Frederick Chambers, Samuel Hunter Christie, Henry Coddington, Henry Thomas Colebrooke, John Cottle, James Cumming, Edmund Robert Daniell, John Frederick Daniell, Martin Davy, George Dollond, George Duckett, John Elliotson, Henry Ellis, William Henry Fitton, Edward Forster, Joseph Henry Green, George Bellas Greenough, Henry Hallam, George Harvey, William Henry, Henry Hennell, Henry Holland, Leonard Horner, Thomas Horsfield, Henry Kater, Henry Bellenden Ker, Philip Parker King, John George Shaw Lefevre, John Lindley, John Augustus Lloyd, John William Lubbock, Charles Lyell, Charles Mackenzie, Gideon Mantell, Herbert Mayo, Roderick Impey Murchison, Whitlock Nicholl, William Hazledine Pepys, William Prout, Thomas Rackett, George Rennie, John Rennie, George Poulett Scrope, Adam Sedgwick, Richard Sheepshanks, Samuel Solly, Richard Horsman Solly, Samuel Reynolds Solly, William Somerville, Charles Stokes, Edward Troughton, Charles Hampden Turner, Edward Turner, James Veitch, John Henry Vivian, Richard Rawlinson Vyvyan, Nathaniel Wallich, Henry Warburton, William Whewell, Richard Willis, and Alexander Luard Wollaston.
53. L. P. Williams, 'The Royal Society and the Founding of the British Association for the Advancement of Science', *Notes and Records of the Royal Society of London*, xvi (1961), 221–233.
54. [Fitton], *A Statement*, pp. 27–28.
55. *The Times*, 1 Dec. 1830.
56. [Fitton], *A Statement*, p. 32.
57. L. G. Wilson, *Charles Lyell. The Years to 1841: The Revolution in Geology* (New Haven, CT, 1972), p. 304.
58. Mrs. [Elizabeth O. Buckland] Gordon, *The Life and Correspondence of William Buckland, D.D., F.R.S.* (London, 1894), pp. 151–152.
59. Alfred Friendly, *Beaufort of the Admiralty: The Life of Sir Francis Beaufort* (London, 1977), p. 287.
60. *Lit. Gaz.*, 723 (1830), 785.
61. *Athenaeum*, 162 (1830), 763.
62. *New Monthly Magazine*, xxxiii (1831), 27.
63. *The Times*, 1 Dec. 1830.
64. *ibid.*
65. *New Monthly Magazine*, xxxiii (1831), 27–28.
66. *The Times*, 3 Dec. 1830.
67. *New Monthly Magazine*, xxxiii (1831), 27.
68. Colvin, *Edgeworth Letters from England*, p. 442.

69. E. C. Curwen (ed.), *The Journal of Gideon Mantell, Surgeon and Geologist* (Oxford, 1940), p. 89.

70. *The Times*, 1 Dec. 1830.

71. Colvin, *Edgeworth Letters from England*, p. 442.

72. *The Times*, 1 Dec. 1830.

73. *Spectator* (London), cxxvii (1830), 992–993.

74. *New Monthly Magazine*, xxxiii (1831), 27–28.

75. *ibid.*

76. A letter from the political economist Richard Jones of Cambridge (1790–1855) to J. F. W. Herschel (RS Herschel Papers) on 9 Jan. 1831 suggests that W. F. Fitton, one of Herschel's campaign managers, is the author of *A Statement . . .* : 'Dr. Fitton (but mind it is a profound secret till *he* tells you) has been preparing a statement of the late struggle for the Chair . . .'. The copy in the Beinecke Rare Book and Manuscript Library at Yale University (New Haven, Connecticut, U.S.A.) was sent to Professor Benjamin Silliman by Charles Babbage through William Vaughan (1742–1850; F.R.S., 1813), a philanthropic merchant with strong scientific and engineering interests and many American connections, who often travelled to New England.

77. [Fitton], *A Statement.*

78. Beinecke Rare Book and Manuscript Library, Yale University, Silliman Papers.

79. Geikie, *Annals of the Royal Society Club*, p. 303.

80. PR, pp. 163-164.

81. Jean Sylvan Bailly, *Histoire de l'astronomie* (Paris, 1775–1787).

82. c.370,20,MSD-3: Augustus De Morgan to WS, n.d. ('Tuesday').

83. Charles Babbage, J. F. W. Herschel and George Peacock, *Examples to the Differential and Integral Calculus* (Cambridge, 1820).

84. c.369,19,MSB-1: Charles Babbage to MS, 23 July 1830.

85. Charles Babbage, 'An Essay Towards the Calculus of Functions' (1815); 'An Essay Towards the Calculus of Functions, Part 2' (1816); 'Observations on the Analogy which Subsists Between the Calculus of Functions and Other Branches of Analysis' (1817).

86. Charles Babbage, 'Observations on the Notation Employed in the Calculus of Functions' (1820).

87. c.369,19,MSB-16: Papers from Charles Babbage.

88. c.369,19,MSB-1: Charles Babbage to MS, 2 Nov. 1830.

89. Bibliothèque nationale NAF 1308 divers 4289 Somerville 98+: MS to Alexis Bouvard, 27 Nov. 1829.

90. PR, pp. 163–164.

91. c.356,6,MSAU-4: 'Mechanism of the Heavens' in MS. notes for autobiography.

92. RS Herschel Papers, 16.327: MS to J. F. W. Herschel, 31 Mar. 1829.

93. RS Herschel Papers, 16.328: MS to J. F. W. Herschel, postmarked 26 Oct. 1829.

94. c.370,20,MSH-3: J. F. W. Herschel to MS, postmarked 5 Feb. 1830.

95. J. F. W. Herschel, 'Sound', *Encyclopaedia Metropolitana*, iv (London, 1830), 810.

96. PR, p. 167.

97. *ibid.*, pp. 168–170.

98. John Pond, *Laplace's System of the World* (London, 1809).

99. John Toplis, *A Treatise Upon Analytical Mechanics: Being the First Book of the Mecanique Celeste of P. S. Laplace* (Nottingham, 1814).

100. Thomas Young, *Elementary Illustrations of the Celestial Mechanism of Laplace (Part the First Comprehending the First Book). With Some Additions Relating to the Motion of Waves and of Sound and Fluids* (London, 1821).

101. Mary Somerville, *The Mechanism of the Heavens* (London, 1831), p. 3.

102. Samuel Smiles, *A Publisher and His Friends: Memoir and Correspondence of the late John Murray* (London, 1891), i. 406.

103. RS Herschel Papers, 16.387: WS to J. F. W. Herschel, 23 Feb. n.y. [1830].

104. Murray Archives: Unpublished letter WS to John Murray, n.d. [May or June 1819].

105. [Mrs.] K. H. Lyell, *Life, Letters, and Journals of Sir Charles Lyell, Bart.* (London, 1881), i. 178.

106. Sir Humphry Davy, *Six Discourses delivered before the Royal Society* (London, 1827).

107. Sir Humphry Davy, *Salmonia or Days of Fly Fishing* (London, 1828).

108. Sir Humphry Davy, *Consolations in Travel: or, the Last Days of a Philosopher* (London, 1830).

109. RS Herschel Papers, 16.387: WS to J. F. W. Herschel, 23 Feb. n.y. [1830].

110. RS Herschel Papers, 16.330: MS to J. F. W. Herschel, 1 Mar. 1830.

111. RS Herschel Papers, 16.331: MS to J. F. W. Herschel, 6 Mar. 1830.

112. RS Herschel Papers, 16.332: J. F. W. Herschel to MS, 9 Mar. 1830.

113. Nathaniel Bowditch, *Mécanique céleste. By the Marquis de la Place &c. Translated with a Commentary by Nathaniel Bowditch* (Boston, MA, 1829).

114. RS Herschel Papers, 16.333: MS to J. F. W. Herschel, n.d. Mar. 1830.

115. RS Herschel Papers, 16.334: J. F. W. Herschel to MS, 31 Mar. 1830.

116. RS Herschel Papers, 16.335 through 16.340: MS to J. F. W. Herschel, 5 Apr. through end of Oct. (undated) 1830.

117. c.373,23,MSBUS-3: John Murray to WS, 2 Aug. 1830.

118. *ibid.*

119. RS Herschel Papers, 16.388: WS to J. F. W. Herschel, 23 Mar. n.y. [1831] (misdated 1830 in RS Catalogue).

120. c.369,19,MSB-1: Charles Babbage to MS, 2 Sept. 1830.

121. c.370,20,MSD-4: J. E. Drinkwater [later J. E. D. Bethune] to MS, 29 July 1831.

122. c.370,20,MSD-4: Thomas Drummond to WS, 9 Dec. 1830.

123. c.369,19,MSB-1: Charles Babbage to MS, 2 Nov. 1830; 26 Feb. 1831; 27 Feb. 1831.

124. c.370,20,MSH-1: Basil Hall to MS, 29 Mar. 1831.

125. PR, p. 168.

126. c.369,19,MSB-13: Henry Brougham to WS, n.d. ('Friday evg').

127. University of St. Andrews Library, James David Forbes Papers: WS to J. D. Forbes, 30 Apr. 1831.

128. J. C. Shairp, P. G. Tait and A. Adams-Reilly, *Life and Letters of James David Forbes, F.R.S.* (London, 1873), p. 69.

129. University of St. Andrews Library, James David Forbes Papers: J. D. Forbes' Notebook of Sketches.

130. c.371,21,MSL-6: Charles Lyell to MS, 3 June 1831.

131. Lyell, *Life of Sir C. Lyell*, i. 324.

132. Curwen, *Journal of G. A. Mantell*, p. 93.

133. Lyell, *Life of Sir C. Lyell*, i. 322.

134. [David Brewster], 'Reflections on the Decline of Science in England . . .', *Quarterly Review*, xliii (1830), 305–342.

135. c.371,21,MSM-5: R. I. Murchison to MS, n.d. ('Sunday Evg') [Oct. or Nov. 1831].

136. Jack Morrell and Arnold Thackray, *Gentlemen of Science* (Oxford, 1981), p. 545.

137. Gerard Moll, *On the Alleged Decline of Science in England by a Foreigner* (London, 1831).

138. *Lit. Gaz.*, 737 (1831), 153–154.

139. *ibid.*

140. RS Herschel Papers, 16.392: WS to J. F. W. Herschel, 14 Sept. 1831.

141. *ibid.*

142. *ibid.*

143. RS Herschel Papers, 16.393: J. F. W. Herschel to WS, 16 Sept. 1831.

144. c.369,19,MSB-13: Henry Brougham to WS, n.d. ('Sunday') [25 Sept. 1831].

145. *Athenaeum*, 166 (1831), 25–27.

146. C. R. Dodd, *A Manual of Dignities, Privileges & Precedence* (London, 1842).

147. Maboth Moseley, *Irascible Genius* (New York, 1964), p. 103.

148. *ibid.*

149. RS Herschel Papers, 16.394: WS to J. F. W. Herschel, 27 Sept. 1831.

150. c.369,19,MSB-13: Henry Brougham to WS, n.d. ('Wednesday') [28 Sept. 1831].

151. RS Herschel Papers, 16.395: WS to J. F. W. Herschel, 29 Sept. 1831.

152. RS Herschel Papers, 16.396: J. F. W. Herschel to WS, 30 Sept. 1831.

153. RS Herschel Papers, 16.398: J. F. W. Herschel to WS, n.d. [4 Oct. 1831].

154. *Athenaeum*, 210 (1831), 726.

155. *Lit. Gaz.*, 772 (1831), 718.

156. Lyell, *Life of Sir. C. Lyell*, i. 322.

157. Somerville, *Mechanism of the Heavens*, p. iii.

158. c.355,5,MSAU-2: pp. 36–37.

159. *ibid.*

160. c.356,6,MSAU-4: MS. notes for autobiography.

161. *Lit. Gaz.*, 778 (1831), 806–807.

162. *Athenaeum*, 221 (1832), 43–44.

163. L. Marchand, *The Athenaeum: A Mirror of Victorian Culture* (Chapel Hill, NC, 1941), pp. 1–12.

164. *Monthly Review*, (NS) i (1832), 133–141.

165. *The Edinburgh New Philosophical Journal* (1832), 376–377.

166. RS Herschel Papers, 16.393: J. F. W. Herschel to WS, 16 Sept. 1831.

167. c.371,21,MSN-1: Macvey Napier to WS, 5 Dec. 1831.

168. *ibid.*

169. [Thomas Galloway], 'Art. I. — Mechanism of the Heavens. By Mrs. Somerville. 8 vo. London: 1830', *Edinburgh Review*, 1v (1832), 1–25.

170. *ibid.*, 1.

171. Lyell, *Life of Sir C. Lyell*, i. 384.

172. *ibid.*, i. 353.

173. [J. F. W. Herschel], 'Art. VIII — 1. Mechanism of the Heavens. By Mrs. Somerville, London, 8 vo., 1832. 2. Mécanique Céleste. By the Marquis de la Place, &c. Translated with a Commentary by Nathaniel Bowditch, LL.D. Volume I. Boston, 1829', *Quarterly Review*, xlvii (1832), 537–559.

174. *ibid.*, 558–559.

175. *ibid.*, 542.

176. *ibid.*, 547.

177. *ibid.*. 558.

178. *ibid.*

179. c.372,22,MSW-1: William Wallace to MS, 23 Dec. 1831.

180. c.369,19,MSB-4: Francis Baily to WS, 17 Dec. 1831.

181. c.371,21,MSI-1: James Ivory to MS, 3 Jan. 1832.

182. J. B. Biot, 'Mécanique céleste, par Mme Sommerville [*sic*]: un volume in 8^0 de 600 pages, avec figures. Londres, 1831.', *Journal des Savans*, Année 1832, 28–32.

183. *Registre des Proces-Verbaux et Rapports des Séances de l'Académie Royale des Sciences.* Année 1832. Tome X, 1re Partie, 23.

184. c.369,19,MSB-8: J. B. Biot to MS, 23 May 1832.

185. c.371,21,MSP-5: S. D. Poisson to MS, 30 May 1832.

186. S. D. Poisson, *Théorie Nouvelle de l'Action Capillaire* (Paris, 1831).

187. c.370,20,MSG-1: J. L. Gay-Lussac to WS, 11 July 1832.

188. Lyell, *Life of Sir C. Lyell*, i. 354.

189. c.371,21,MSP-4: Julius Plücker to Leonard Horner, 11 Jan. 1832.

190. c.372,22,MSW-2: William Whewell to MS, 2 Dec. 1831.

191. Lyell, *Life of Sir C. Lyell*, i. 368.

192. *ibid.*, i. 352–353.

193. *ibid.*, i. 368.

194. *ibid.*, 1. 371.

195. Mary Somerville, *Preliminary Dissertation to the 'Mechanism of the Heavens'* (London, 1831).

196. Lyell, *Life of Sir C. Lyell*, i. 368.
197. c.375,25,MSDIP-2: WS to Mrs. Janet Somerville Pringle Elliot, 29 Feb. 1832.
198. c.370,20,MSE-1: Maria Edgeworth to MS, 31 Mar. 1832.
199. Mary Somerville, *A Preliminary Dissertation on the Mechanism of the Heavens* (Philadelphia, 1832).
200. PR, pp. 172–173.
201. c.371,21,MSP-1: George Peacock to MS, 14 Feb. 1832.
202. Lyell, *Life of Sir C. Lyell*, i. 373.
203. *ibid.*, i. 373–374.
204. c.355,5,MSAU-3: p. 34.
205. c.356,6,MSAU-4: MS. notes for autobiography.
206. c.375,25,MSDIP-2: J. G. Children to MS, 23 Dec. 1831.
207. Lyell, *Life of Sir C. Lyell*, i. 371.
208. c.375,25,MSDIP-2: WS to Mrs Janet Somerville Pringle Elliot, 29 Feb. 1832.
209. c.375,25,MSDIP-2: J. G. Children to WS, 19 Feb. 1832.
210. c.375,25,MSDIP-2: List of Subscribers to Chantrey Bust of MS.
211. Lyell, *Life of Sir C. Lyell*, i. 373.
212. c.356,6,MSAU: p. 112.
213. c.370,20,MSH-4: Margaret Herschel to MS, 2 Apr. 1832.
214. c.371,21,MSK-1: Mary Frances Kater (Mrs. Henry) to MS, 12 Apr. 1832.
215. c.369,19,MSA-1: G. B. Airy to WS, 30 Mar. 1832.
216. c.372,22,MSS-4: Adam Sedgwick to WS, 3 Apr. 1832.
217. c.372,22,MSS-4: Adam Sedgwick to WS, 5 Apr. 1832.
218. R. P. Graves, *Life of Sir William Rowan Hamilton* (Dublin, 1882), i. 553.
219. Private communication, I. M. McCabe, Librarian, Royal Institution.
220. Gordon, *Life of William Buckland*, p. 123.
221. Magdalen College, Oxford, Daubeny Letter Book: Charles Babbage to C. G. B. Daubney, 28 Apr. 1832.
222. Wilson, *Lyell*, p.353.
223. Lyell, *Life of Sir C. Lyell*, i. 381.
224. *ibid.*, i. 382.
225. Trinity College (Cambridge) Library, Whewell Papers: MS to William Whewell, 4 May 1832.
226. Lyell, *Life of Sir C. Lyell*, i. 383.
227. *ibid.*
228. c.365,15,MSIF-34: Martha S to WG, 13 May 1833.
229. Lyell, *Life of Sir C. Lyell*, i. 397.
230. NLS 2213.f.73: MS to Mrs. Leonard Horner, 12 June 1832.
231. Gordon, *Life of William Buckland*, p. 123.

Chapter 6 — The Second Stay Abroad
pages 95–121

1. R. J. Morris, *Cholera 1832: The Social Response to an Epidemic* (New York, 1976), p. 75.
2. c.357,7,MSFP-5: MS to Lady Fairfax, 29 Sept. 1832.
3. Lyell, *Life of Sir C. Lyell*, i. 313.
4. J. F. Cooper, *A Residence in France, with an Excursion up the Rhine and a Second Visit to Switzerland* (Paris, 1836).
5. b.205,29,MSFP$_2$-60: WS to Lady Fairfax, 2 Nov. 1832.
6. b.205,29,MSFP$_2$-60: WG to Lady Fairfax, 23 Oct. 1832.
7. c.369,19,MSB-10: Alexis Bouvard to WS, 25 Sept. 1832.
8. c.363,13,MSIF-13: WS to WG, n.d. [22 Oct. 1832].
9. c.361,11,MSIF-1: MS to WG, n.d. ('Sunday') [8 Oct. 1832; later misdated by WG '1 Oct'].
10. V. Y. Bowditch, *Life and Correspondence of Henry Ingersoll Bowditch* (Boston, MA, 1902), ii. 131–132.

11. c.363,13,MSIF-13: WS to WG, n.d. [22 Oct. 1832].

12. *ibid.*

13. c.365,15,MSIF-34: Mary CS to WG, 31 Oct. 1832.

14. c.365,15,MSIF-34: Mary CS to WG, 8 Feb. 1833.

15. c.363,13,MSIF-13: WS to WG, 16 Nov. 1832.

16. c.370,20,MSD-4: Charles Dupin to WS, 10 Aug. 1820.

17. c.362,12,MSIF-12: MS to WS, 12 Apr. 1833.

18. PR, pp. 178–179.

19. c.358,8,MSFP-17: Anne L. Napier (Mrs. Richard) to MS, 27 May n.y. [1832].

20. J. C. Maxwell, 'Grove's Correlation of Physical Forces', *The Scientific Papers of J. Clerk Maxwell* (Cambridge, 1890), ii. 401.

21. Mary Somerville, *On the Connexion of the Physical Sciences* (London, 1834).

22. PR, pp. 178–179.

23. c.369,19,MSB-12: W. J. Broderip to MS, 15 Mar. 1832.

24. W. Moorcroft, 'Asiatic Researches, xii', *Asiatic Society's Transactions*, i. 375–534.

25. John Gould, *A Century of Birds from the Himalaya Mountains* (London, 1832).

26. c.369,19,MSB-12: W. J. Broderip to MS, 15 Mar. 1832.

27. OCPS (1834), pp. 275–277.

28. Trinity College (Cambridge) Library, Whewell Papers: MS to William Whewell, 4 May 1832.

29. NLS 2213.f.73: MS to Mrs. Leonard Horner, 12 June 1832.

30. PR, pp. 183–184.

31. *ibid.*

32. c.369,19,MSB-10: Alexis Bouvard to WS, 3 Nov. 1832.

33. c.369,19,MSB-10: Alexis Bouvard to WS, 26 Nov. 1832.

34. c.361,11,MSIF-1: MS to WG, 12 Apr. 1833.

35. c.362,12,MSIF-12: MS to WS, 12 Apr. 1833.

36. OCPS (1834), pp. 364–365.

37. *ibid.*, pp. 219–220.

38. *ibid.*, p. 135.

39. *ibid.*, pp. 267–268.

40. *ibid.*. pp. 341–344.

41. PR, p. 185.

42. c.369,19,MSB-8: J. B. Biot to MS, 5 May 1833.

43. OCPS (1834), pp. 214–216.

44. c.369,19,MSB-8: J. B. Biot to MS, 5 May 1833.

45. c.370,20,MSG-1: Josephine Gay-Lussac to MS, 23 Feb. 1819.

46. c.361,11,MSIF-1: MS to WG, 16 Dec. 1832.

47. c.365,15,MSIF-34: Mary CS to WG, 1 Jan. 1833.

48. PR, p. 187.

49. *ibid.*

50. c.362,12,MSIF-8: MS to Martha S, 11 Oct. n.y. [1835].

51. Mary Somerville, *On the Connexion of the Physical Sciences* (2nd edn., London, 1835), p. 130.

52. *ibid.*, p. 311.

53. c.371,21,MSL-2: Mme Laplace to MS, 6 Feb. 1833.

54. c.361,11,MSIF-1: MS to WG, 28 Feb. 1833.

55. c.371,21,MSM-1: François Magendie to MS, n.d. [shortly after 5 Mar. 1827].

56. PR, pp. 192–193.

57. c.355,5,MSAU-3: pp. 88–89.

58. c.361,11,MSIF-1: MS to WG, 21 Jan. 1833.

59. *ibid.*

60. *ibid.*

61. c.361,11,MSIF-1: MS to WG, 28 Feb. 1833; 24 Apr. 1833.

62. c.371,21,MSL-4: J. W. Lubbock to MS, n.d.; n.d. [Dec. 1831]; 13 Feb. n.y. [1832]; 13 Mar. n.y. [1832 or 1833].

63. RS Lubbock Papers, 36.s.292: MS to J. W. Lubbock, 23 Feb. n.y. [1831 or 1832].

64. c.371,21,MSL-4: J. W. Lubbock to MS, 28 Dec. 1832.

65. c.371,21,MSP-5: S. D. Poisson to MS, 30 May 1832.

66. c.371,21,MSP-5: S. D. Poisson to MS, 28 Nov. 1832.

67. c.355,5,MSAU-2: p. 141.

68. c.371,21,MSP-5: S. D. Poisson to MS, 30 May 1832.

69. c.361,11,MSIF-1: MS to WG, n.d. [7] Oct. 1832.

70. PR, pp. 201–202.

71. Maurice Crosland, 'Development of the Professional Career in Science in France' in Maurice Crosland (ed.), *The Emergence of Science in Western Europe* (New York, 1976), pp. 139–159.

72. PR, p. 186.

73. RS Certificate Ledger, 1820–1829: William Frederick Milne Edwards. Elected 26 Feb. 1829.

74. c.355,5,MSAU-2: p. 143.

75. W. F. M. Edwards, *Influence of Physical Agents on Life* (trans. by Hodgkin and Fisher) (London, 1832).

76. PR, p. 187.

77. *ibid.*, p. 186.

78. Crosland, *Society of Arcueil*, p. 323.

79. OCPS (1834), pp. 334–335.

80. *ibid.*, p. 326.

81. OCPS (1835), pp. 307–308.

82. *ibid.*, p. 326.

83. *ibid.*, p. 363.

84. Crosland, *Society of Arcueil*, p. 95.

85. c.362,12,MSIF-12: MS to WS, 12 Apr. 1833.

86. c.362,12,MSIF-12: MS to WS, 27 May 1833.

87. Beinecke Rare Book and Manuscript Library, Yale University, uncatalogued letter; MS to 'Mlle de Vaucel', 14 Jan. 1828.

88. Mrs. [Sarah W. B.] Lee, *Memoirs of Baron Cuvier* (London, 1833), p. 308.

89. 'Cuvier', *Dictionary of Scientific Biography* (New York, 1971), iii. 521–528.

90. c.365,15,MSIF-34: Mary CS to WG, 8 Nov. 1832.

91. c.365,15,MSIF-34: Mary CS to WG, 15–16 Nov. 1832.

92. c.365,15,MSIF-32: Martha S to WG, 20 Nov. 1832.

93. c.355,5,MSAU-2: p. 143.

94. D. I. Duveen and H. S. Klickstein, *A Bibliography of the Works of A. L. Lavoisier* (London, 1954), p. 3.

95. c.365,15,MSIF-32: Martha S to WG, 20 Nov. 1832.

96. PR, pp. 188–189.

97. c.365,15,MSIF-32: Martha S to WG, 28 Feb. 1833.

98. c.361,11,MSIF-1: MS to WG, 3 May 1833.

99. c.365,15,MSIF-34: Mary CS to WG, n.d. ('last week in March') [1833]; 31 Mar. 1833.

100. c.361,11,MSIF-1: MS to WG, 24 Apr. n.y. [1833]; 8 May 1833.

101. c.365,15,MSIF-34: Mary CS to WG, n.d. ('last week in March') [1833].

102. c.361,11,MSIF-1: MS to WG, 10 June 1833.

103. PR, p. 194.

104. Bowditch, *Life of H. I. Bowditch*, i. 10, 21–22.

105. b.205,29,MSFP$_2$-60: WG to Lady Fairfax, 23 Oct. 1832.

106. c.361,11,MSIF-1: MS to WG, 23 Nov. 1832.

107. c.371,21,MSL-2: Mme Laplace to MS, 9 Jan. 1833.

108. c.369,19,MSB-10: Eugene Bouvard to MS, 9 May 1833.

109. J. W. Lubbock, 'On the Calculation of Annuities, and on Some Questions on the Theory of Chances', *Transactions of the Cambridge Philosophical Society*, iii (1830), 141–155. This is Lubbock's first published paper.

110. Between 1831 and 1835, Lubbock published three papers on the theory of the moon in the *Philosophical Transactions*. See *Abstracts of . . . the Philosophical Transactions . . .* , iii, (1831) 75; (1833) 253; (1834) 270.

111. In 1831 Lubbock published the first two of his many papers on the tides; 'On the Tides on the Coast of Great Britain', *Philosophical Magazine*, ix (1833), 333–335 and 'On the Tides in the Port of London', *Philosophical Transactions*, cxxii (1831), 379–416.

112. RS Lubbock Papers, 36.s.289: MS to J. W. Lubbock, n.d. [1830]; s.290: MS to J. W. Lubbock, n.d. [1830]; s.291: MS to J. W. Lubbock, n.d. [1830]; s.292: MS to J. W. Lubbock, 23 Feb. n.y. [1831 or 1832].

113. c.371,15,MSL-5: J. W. Lubbock to MS, n.d. ('With the best consideration that I can give to it I think you are right . . .') [pre-April 1830]; 13 Mar. n.y. [1831]; n.d. ('When I was in Paris lately M. Poisson . . .') [Dec. 1831]; 13 Feb. n.y. [1832]; 28 Dec. 1832.

114. RS Lubbock Papers, 36.s.284: MS to J. W. Lubbock, n.d. ('Wednesday') [1830]; s.285: 3 July n.y. [1830]; s.286: 25 Aug. n.y. [1830]; s.287: n.d. ('Saturday') [1830]; s.288: n.d. ('Saturday') [1830].

115. c.371,21,MSL-5: J. W. Lubbock to MS, n.d. ('At last I have the pleasure of sending you . . .') [pre-April 1830]; 2 Oct. n.y. [1831].

116. c.361,11,MSIF-1: MS to WG, 23 Nov. 1832.

117. c.371,21,MSL-5: J. W. Lubbock to MS, 28 Dec. 1832.

118. Robert Robson, 'Trinity College in the Age of Peel', in Robert Robson (ed.), *Ideas and Institutions of Victorian Britain* (London, 1967), p. 329.

119. *ibid.*, p. 315.

120. c.365,15,MSIF-34: Mary CS to WG, n.d. ('First week in May') [probably 10 May 1833].

121. RS Certificate Ledger, 1832–1833: Woronzow Greig. Elected 7 February 1833.

122. Bowditch, *Life of H. I. Bowditch*, i. 30–32.

123. *ibid.*, pp. 50–54.

124. c.365,15,MSIF-32: Martha S to WG, 2 Feb. 1833.

125. c.365,15,MSIF-34: Mary CS to WG, 8 Feb. 1833.

126. c.370,20,MSC-5: Clot Bey to WS, 2 Jan. 1834.

127. J. P. T. Bury (ed.), *Romilly's Cambridge Diary 1832–42* (Cambridge, 1967), p. 29.

128. c.361,11,MSIF-1: MS to WG, 28 Feb. 1833.

129. c.370,20,MSC-5: Clot Bey to WS, 2 Jan. 1834.

130. S. F. Cannon, *Science in Culture* (New York, 1978), p. 186.

131. *Testimonials in Favour of James D. Forbes, F.R.S.S L. & E., F. G.S., Member of the Royal Geographical Society, of the Society of Arts for Scotland, and Hon. Member of the Yorkshire Philosophical Society as a Candidate for the Chair of Natural Philosophy in the University of Edinburgh*. 1832. [Privately printed, 72 pp.].

132. University of St. Andrews Library, James David Forbes Papers: MS to J. D. Forbes, n.d. [May 1833].

133. c.365,15,MSIF-32: Martha S to WG, 3–4 Dec. 1832.

134. c.370,20,MSF-2: J. D. Forbes to MS, 19 May 1833.

135. c.365,15,MSIF-32: Martha S to WG, 8 Nov. n.y. [1832].

136. c.365,15,MSIF-32: Martha S to WG, n.d. [7] Feb. 1833.

137. PR, pp. 190–191.

138. c.365,15,MSIF-34: Mary CS to WG, 8 Feb. 1833.

139. c.362,12,MSIF-12: MS to Mary CS, n.d. ('Tuesday') [28] July 1833.

140. Somerville College (Oxford), Somerville Memorabilia: MS to Prince Koslofski, 5 May 1837.

141. c.361,11,MSIF-1: MS to WG, 14 June 1833.

142. c.365,15,MSIF-32: Martha S to WG, 3 May 1833.

143. c.365,15,MSIF-32: Martha S to WG, 19 June 1833.

144. c.371,21,MSI-1: James Ivory to MS, 9 Jan. 1833.

145. c.369,19,MSB-4: Francis Baily to WS, 3 Feb. 1833.
146. c.370,20,MSD-1: T. S. Davies to MS, 28 June 1832.
147. c.370,20,MSD-1: T. S. Davies to MS, 24 Jan. 1833.
148. RS Certificate Ledger, 1832–1833: T. S. Davies. Elected 18 Apr. 1833.
149. c.372,22,MSW-2: William Whewell to MS, 3 Apr. 1833.
150. William Whewell, *Astronomy and General Physics, Considered with Reference to Natural Theology* (London, 1833).
151. c.372,22,MSW-2: William Whewell to MS, 3 Apr. 1833.
152. c.361,11,MSIF-1: MS to WG, 24 Apr. 1833.
153. c.361,11,MSIF-1: MS to WG, 14 June 1833.
154. c.371,21,MSM-5: Charlotte Murchison (Mrs. R. I.) to MS, 6 May 1833.
155. William Sotheby, *Lines Suggested by the Third Meeting of the British Association for the Advancement of Science, Held at Cambridge, in June, 1833, by the Late William Sotheby, Esq. F.R.S. &c With a Short Memoir of his Life* (London, 1834).
156. *ibid.*, p. 18.
157. c.365,15,MSIF-34: Mary CS to WG, n.d. ('Thursday') [7 Mar. 1833].
158. c.365,15,MSIF-32: Martha S to WG, 22 Mar. 1833.
159. c.373,23,MSBUS-1: Murray Accounts, 20 Mar. 1833.
160. c.373,23,MSBUS-1: Murray Accounts, 'Mar. 1833'.
161. c.373,23,MSBUS-3: John Murray to WS, 21 Mar. 1833.
162. Murray Archives: MS to John Murray, 2 Apr. 1833.
163. c.365,15,MSIF-34: Mary CS to WG, 8 Feb. 1833.
164. c.365,15,MSIF-32: Martha S to WG, 3–4 Dec. 1832.
165. c.365,15,MSIF-34: Mary CS to WG, n.d. ('Thursday') [7 Mar. 1833].
166. c.365,15,MSIF-32: Martha S to WG, 22 Mar. 1833.
167. Wilson, *Lyell*, p. 341.
168. c.365,15,MSIF-32: Martha S to WG, 3–4 Dec. 1832.
169. J. F. W. Herschel, *A Preliminary Discourse on the Study of Natural Philosophy* (London, 1832).
170. c.361,11,MSIF-1: MS to WG, 28 Feb. 1833.
171. c.365,15,MSIF-34: Mary CS to WG, n.d. ('Thursday') [7 Mar. 1833].
172. Murray Archives: MS to John Murray, 2 Apr. 1833.
173. c.361,11,MSIF-1: MS to WG, 24 Apr. 1833.
174. c.361,11,MSIF-1: MS to WG, 23 May 1833.
175. c.365,15,MSIF-32: Martha S to WG, 19 May 1833.
176. PR, p. 193.
177. c.362,12,MSIF-12: MS to WS, 31 May 1833.
178. *ibid.*
179. c.361,11,MSIF-1: MS to WG, 14 June 1833.
180. J. F. W. Herschel, *Treatise on Astronomy* (London, 1833).
181. c.361,11,MSIF-1: MS to WG, 14 June 1833.
182. c.362,12,MSIF-12: MS to WS, 2 Apr. 1833.
183. c.362,12,MSIF-12: MS to WS, 31 May 1833.
184. c.371,21,MSM-5: Charlotte Murchison (Mrs. R. I.) to MS, 6 May 1833.
185. C. G. T. Dean, *The Royal Hospital at Chelsea* (London, 1950), p. 274.
186. c.364,14,MSIF-21: WG to WS, 19 Sept. 1833.
187. PR, p. 196.
188. c.365,15,MSIF-32: Martha S to WG, 31 Mar. 1833.
189. c.362,12,MSIF-12: MS to WS, 18 Aug. 1833.
190. c.364,14,MSIF-21: WG to WS, 19 Sept. 1833.
191. c.362,12,MSIF-12: MS to WS, 14 June 1833.

Chapter 7 — On the Connexion of the Physical Sciences
pages 123–150

1. *The Oxford English Dictionary* (Oxford, 1933), vii, 808.

2. PR, pp. 178–179.
3. John Robison, 'Physics', *Encyclopaedia Britannica* (3rd edn., Edinburgh, 1797), xvii. 637–660.
4. PR, p. 132.
5. Thomas Young, *A Course of Lectures on Natural Philosophy and the Mechanical Arts* (London, 1807), ii. 487.
6. John Playfair, *Outlines of Natural Philosophy, being the heads of lectures delivered at the University of Edinburgh* (3rd edn., Edinburgh, 1819), i. v.
7. *ibid.*, p. 1.
8. *ibid.*, p. 16.
9. William Whewell, *An Elementary Treatise on Mechanics* (Cambridge, 1819).
10. William Whewell, *Syllabus of an Elementary Treatise on Mechanics* (1821); *A Treatise on Dynamics* (1832); *Analytical Statics: A Supplement to the 4th Edition of An Elementary Treatise on Mechanics* (1833) (Cambridge, dates as specified).
11. Trinity College (Cambridge) Library, Add. MS. a. 212[125]: MS to William Whewell, 7 June n.y. [1832].
12. Playfair, *Outlines*, i. 341.
13. Crosbie Smith, '"Mechanical Philosophy" and the Emergence of Physics in Britain: 1800–1850', *Annals of Science*, xxxiii (1976), 13.
14. J. F. W. Herschel, *Preliminary Discourse*, p. 93.
15. *ibid.*, pp. 223–224.
16. *ibid.*
17. *ibid.*, pp. 344–346.
18. PR, p. 178.
19. Thomas Thomson, *System of Chemistry* (7th edn., Edinburgh, 1831).
20. Maurice Crosland, *Gay-Lussac: Scientist and Bourgeois* (Cambridge, 1978), pp. 117–128.
21. O. J. R. Howarth, *The British Association for the Advancement of Science: A Retrospect 1831–1931* (2nd edn., London, 1931), p. 19.
22. *ibid.*, p. 24.
23. Wilson, *Lyell*, p. 435.
24. Charles Babbage, *The Ninth Bridgewater Treatise: A Fragment* (London, 1837).
25. J. W. Clark and T. McK. Hughes, *Life and Letters of the Reverend Adam Sedgwick* (Cambridge, 1890), i. 483–484.
26. Wilson, *Lyell*, pp. 461–462.
27. PR, pp. 212–214.
28. c.370,20,MSH-4: Margaret Herschel to MS, 3 Nov. 1833.
29. c.369,19,MSB-1: Charles Babbage to MS, 14 Dec. 1833.
30. c.372,22,MSW-2: William Whewell to WS, 30 Aug. 1833.
31. c.372,22,MSW-2: William Whewell to WS, n.d. [Sept. 1833].
32. c.372,22,MSW-2: William Whewell to WS, 22 Nov. 1833.
33. *ibid.*
34. c.369,19,MSB-13: Henry Brougham to WS, n.d. ('Saturday') [12 Sept. 1833].
35. c.369,19,MSB-13: Henry Brougham to MS, n.d. ('Saturday') [Sept. or Oct. 1833].
36. OCPS (1834), p. 185.
37. c.375,25,MSDIP-9: Draft of dedication to Queen Adelaide.
38. c.370,20,MSF-2: J. D. Forbes to WS, 20 Aug. 1833.
39. c.370,20,MSF-2: J. D. Forbes to WS, 21 Oct. 1833.
40. c.370,20,MSH-5: Henry Holland to MS, n.d. ('Monday Evg'; 'Friday Evg') [1833].
41. c.370,20,MSH-5: Henry Holland to MS, n.d. ('Friday Evg') [1833].
42. c.370,20,MSH-5: Henry Holland to MS, 26 Oct. n.y. [1833].
43. [Henry Holland], 'The Bridgewater Treatises [Whewell, Kidd, Bell and Chalmers]. The Universe and its Author', *Quarterly Review*, L (1833), 1–34.
44. c.370,20,MSH-5: Henry Holland to MS, 26 Oct. n.y. [1833].
45. [Holland], *Quarterly Review*, L (1833), 11.

46. *Lit. Gaz.* 894 (1834), 183; 896 (1834), 215; 897 (1834), 232; 935 (1834), 855; 936 (1834), 877; 1003 (1836), 286; 1029 (1836), 751.
47. c.371,21,MSL-6: Charles Lyell to MS, 23 Dec. 1833.
48. Wilson, *Lyell,* pp. 385–386.
49. c.371,21,MSL-6: Charles Lyell to MS, 23 Dec. 1833.
50. c.371,21,MSL-6: Charles Lyell to MS, n.d. ('Monday') [30 Dec. 1833 or 6 or 13 Jan. 1834]; 24 Feb. 1834; 27 Feb. n.y. [1834].
51. c.370,20,MSF-1: Michael Faraday to WS, Nov. 1833.
52. Michael Faraday, *Experimental Researches in Electricity* (London, 1914) [Everyman's Library edition], pp. 47, 84, 111, 171.
53. OCPS (1835), pp. 130–131.
54. c.370,20,MSF-1: Michael Faraday to WS, Nov. 1833.
55. OCPS (1834), p. 290.
56. c.370,20,MSF-1: Michael Faraday to WS, Nov. 1833.
57. OCPS (1834), p. 292.
58. c.363,23,MSBUS-1: Murray Statement, Jan.-June 1834.
59. *Lit. Gaz.,* 891 (1834), 127.
60. *ibid.,* 894 (1834), 183.
61. *Athenaeum,* 333 (1834), 216.
62. *Lit. Gaz.,* 896 (1834), 215.
63. *ibid.,* 897 (1834), 232.
64. *Athenaeum,* 340 (1834), 343.
65. c.361,11,MSIF-1: MS to WG, 6 Mar. 1834.
66. c.373,23,MSBUS-1: Murray Statement, June 1834.
67. Robert Southey, *The Doctor* (London, 1834).
68. Bury, *Romilly's Cambridge Diary,* p. 53.
69. *Lit. Gaz.,* 894 (1834), 173.
70. *Athenaeum,* 333 (1834), 202–203.
71. *ibid.*
72. *Mechanics Magazine,* xx, 555 (1834), 442–447.
73. c.370,20,MSF-2: J. D. Forbes to WS, 21 Oct. 1833.
74. c.370,20,MSH-5: Henry Holland to MS, 28 Dec. n.y. [1833].
75. c.371,21,MSL-6: Charles Lyell to MS, n.d. ('Monday') [30 Dec. 1833].
76. [William Whewell], 'Art. III. — On the Connexion of the Physical Sciences. By Mrs. Somerville', *Quarterly Review,* li (1834), 54–68.
77. *ibid.,* 68.
78. c.371,21,MSN-1: Macvey Napier to WS, 5 Dec. 1833.
79. c.371,21,MSN-1: Macvey Napier to WS, 15 Dec. 1833.
80. c.369,19,MSB-1: David Brewster to MS, 10 Jan. 1834.
81. [David Brewster], 'Art. VIII. — On the Connexion of the Physical Sciences. By Mrs. Somerville', *Edinburgh Review,* lix (1834), 154–171.
82. *The British Critic, Quarterly Theological Review and Ecclesiastical Record,* xvi (1834), 123–132.
83. Mrs. Somerville, *On the Connexion of the Physical Sciences* (Philadelphia, 1834).
84. *The American Journal of Science and the Arts,* xxvii (1835), 396.
85. 'Electro Magnetism. History of Davenport's Invention of the Application of Electro-Magnetism to Machinery; with remarks on the same from the *American Journal of Science and the Arts,* by Prof. Silliman. Also, extracts from other public journals, and information on electricity, galvanism, electro-magnetism, etc. by Mrs. Somerville.' (New York, 1837).
86. *Lit. Gaz.,* 1063 (1837), 356.
87. c.371,21,MSM-3: Harriet Martineau to MS, 1 Nov. 1836.
88. c.371,21,MSM-3: Harriet Martineau to MS, n.d. ('Tuesday evg').
89. c.355,5,MSAU-3: p. 42 in added sheets.
90. c.371,21,MSP-3: Thomas Phillips to MS, 3 Feb. 1834.
91. Murray Archives: MS to John Murray, 4 Feb. 1834.

92. c.371,21,MSP-3: Thomas Phillips to MS, 22 June n.y. [1834].

93. c.361,11,MSIF-1: MS to WG, 6 Mar. 1834.

94. *Athenaeum*, 342 (1834), 379.

95. *Lit. Gaz.*, 904 (1834), 345.

96. c.361,11,MSIF-1: MS to WG, 6 Mar. 1834.

97. c.375,25,MSDIP-9: Sir Herbert Taylor to Lord John Russell, 25 Mar. 1834.

98. c.372,22 MSR-2: Lord John Russell to WS, 25 Mar. n.y. [1834].

99. c.375,25,MSDIP-9: James Hudson to WS, 2 Apr. 1834.

100. c.375,25,MSDIP-6: Diploma, Société de Physique et d'Histoire Naturelle de Genève, 3 Apr. 1834.

101. c.375,25,MSDIP-3: Diploma, Royal Irish Academy, 26 May 1834.

102. c.371,21,MSM-2: Jane Marcet to MS, 6 Apr. 1834.

103. c.375,25,MSDIP-3: William Rowan Hamilton to MS, 27 May 1834.

104. c.371,21,MSM-2: Jane Marcet to MS, 26 Oct. 1833.

105. c.360,10,MSFP-54: Sir A. S. Greig to MS, 10/22 Sept. 1834.

106. Unpublished letter owned by Mrs. C. Colvin, Oxford: Maria Edgeworth to Fanny Wilson, 3 May 1835.

107. *ibid.*

108. c.369,19,MSB-5: Francis Beaufort to MS, 27 Oct. 1834.

109. PR, p. 176.

110. *The Times*, 2 Jan. 1835.

111. c.369,19,MSB-5: William Potter of Liverpool to MS, 21 Mar. 1835.

112. *ibid.*

113. PR, p. 176.

114. M. J. Cullen, *The Statistical Movement in Early Victorian Britain: The Foundations of Empirical Social Research* (New York, 1975), p. 79.

115. c.369,19,MSB-1: Charles Babbage to WS, n.d. [week of 9 Mar. 1834].

116. Cullen, p. 79.

117. *Annals of the Royal Statistical Society 1834–1934* (London, 1934), pp. 16–17.

118. *Athenaeum*, 369 (1834), 859.

119. c.356,6,MSAU-4: 'Poisson' in noted for autobiography.

120. PR, p. 202.

121. *ibid.*

122. c.369,19,MSB-1: Charles Babbage to MS, 16 July 1834.

123. c.371,21,MSL-4: J. W. Lubbock to WS, 26 Aug. 1834.

124. Charles Wheatstone, 'On the Figures obtained by strewing Sand on Vibrating Surfaces, commonly called Acoustic Figures', *Philosophical Transactions*, 1833, 593–633.

125. c.372,22,MSS-6: W. H. Smyth to MS, 24 Apr. 1834; 6 Aug. 1834; 5 Sept. 1834; 12 Sept. 1834.

126. OCPS (1835), pp. 385–392, 400.

127. c.370,20,MSF-1: Michael Faraday to MS, 1 Mar. 1834.

128. Faraday, *Experimental Researches in Electricity*, pp. 172–232.

129. Williams, *Faraday Correspondence*, i. 264–272.

130. c.370,20,MSF-1: Michael Faraday to MS, Aug. 1834.

131. OCPS (1835), pp. 124–140.

132. c.371,21,MSJ-1: Francis Jeffrey to MS, 16 Aug. 1834.

133. *Athenaeum*, 922 (1834), 637.

134. c.361,11,MSIF-1: MS to WG, 6 Oct. 1834.

135. c.370,20,MSF-1: Michael Faraday to MS, 10 Oct. 1834.

136. OCPS (1835), pp. 419–425.

137. [Brewster], *Edinburgh Review*, lix (1834), 160.

138. *Lit. Gaz.*, 935 (1834), 855.

139. c.373,23,MSBUS-1: Murray Statement, Dec. 1834–June 1835.

140. c.375,25,MSDIP-2: J. G. Children, Secretary RS, Receipt, 24 July 1835.

141. c.375,25,MSDIP-3: A. McFarlane, London Mechanics Institution, to MS, 19 Oct. 1835.

142. *Lit. Gaz.*, 935 (1834), 855; 936 (1834), 877; 944 (1835), 127; 945 (1835), 143; 948 (1835), 192; 951 (1835), 239.

143. *Athenaeum*, 374 (1834), 942; 382 (1835), 160; 386 (1835), 231.

144. c.373,23,MSBUS-1: Murray Statement, Dec. 1834–June 1835.

145. Harriet Martineau, *Biographical Sketches*, p. 388.

146. The geologist and mineralogist Warington Wilkinson Smyth (1817–1890; F. R. S., 1858); the astronomer Charles Piazzi Smyth (1819–1900; F. R. S., 1857; resigned 1874); Georgiana Rossetta Smyth, who married in 1858 William Henry Flower (1831–1899; F. R. S., 1864), Director of the Natural History Museum, London; and Henrietta Grace Smyth, who married in 1846, as his second wife, Professor Baden Powell. A third Smyth son became Gen. Sir Henry Augustus Smyth (1825–1906).

147. c.369,19,MSB-5: Henrietta Beaufort to MS, 12 Dec. 1836.

148. Lyell, *Life of Sir C. Lyell*, i. 381.

149. Trinity College (Cambridge) Library, Whewell Papers: MS to William Whewell, 4 May 1832.

150. Beinecke Rare Book and Manuscript Library, Yale University: uncatalogued letter, MS to Mlle DuVaucel [Davaucelle], 14 Jan. 1828.

151. c.367,17,MSBY-1: Lady Noel Byron to MS, 4 Mar. n.y. [1829?].

152. E. C. Mayne, *Life and Letters of Anne Isabella, Lady Noel Byron* (London, 1929), p. 324.

153. *ibid.*

154. b.205,29,MSIF$_2$-40: Fragmentary memoir of Lady Lovelace by WG.

155. c.367,17,MSBY-1: Lady Noel Byron to MS, 4 Mar. n.y. [1829?].

156. G. B. Ticknor, *Life, Letters and Journals of George Ticknor* (Boston, MA, 1876), i. 410–411.

157. c.367,17,MSBY-2: Ada Augusta Byron to MS, n.d. ('Monday morning 8 Nov') [1830].

158. b.205,29,MSIF$_2$-40: Fragmentary memoir of Lady Lovelace by WG.

159. c.367,17,MSBY-2: Ada Augusta Byron to MS, 15 Nov. 1834.

160. c.362,12,MSIF-8: MS to Martha S, 25 Dec. 1835.

161. Mayne, *Life of Lady Noel Byron*, pp. 477–478.

Chapter 8 — The Civil List and Mary Somerville
pages 151–162

1.   c.357,7,MSFP-2: Lady Fairfax to MS, 29 Dec. 1830.

2.   c.361,11,MSIF-1: MS to WG, 21 Jan. 1833.

3.   Clark and Hughes, *Life of Sedgwick*, i. 432–433.

4.   C. S. Parker, *Sir Robert Peel* (London, 1899), i. 100.

5.   I Will. IV. c.25.

6.   Babbage, *Reflections on the Decline of Science*, pp. 14–28.

7.   *The Times*, 19 Nov. 1831.

8.   *Athenaeum*, 226 (1832), 130.

9.   Patterson, *John Dalton*, pp. 233–234.

10. *The Times*, 5 Aug. 1830.

11. Parker, *Peel*, ii. 304.

12. L. J. Jennings (ed.), *The Correspondence and Diaries of the Late Rt. Hon. J.W. Croker* (London, 1885), ii. 257–258.

13. *ibid.*, ii. 259.

14. *ibid.*

15. Andrew Lang, *Life and Letters of John Gibson Lockhart* (London, 1897), ii. 216.

16. Wilfred Airy (ed.), *Autobiography of Sir George Biddle Airy* (Cambridge, 1896), pp. 104–105.

17. *ibid.*, p. 108.

18. *ibid.*

19. *ibid.*, p. 105.

20. c.375,25,MSDIP-3: Augustus De Morgan, Secretary of the R.A.S., to MS, 13 Feb. 1835.

21. Mrs. John Herschel, *Memoir and Correspondence of Caroline Herschel* (London, 1876), p. 271.
22. *ibid.*, p. 274.
23. *The Times*, 4 Mar. 1835.
24. *Lit. Gaz.*, 946 (1835), 156.
25. c.375,25,MSDIP-9: Sir John Conroy to WS, 25 Feb. 1835.
26. c.361,11,MSIF-1: MS to WG, 16 Mar. 1835.
27. PR, p. 203.
28. c.367,17,MSBY-2: Ada Augusta Byron to MS, 12 Mar. 1835.
29. PR, p. 203.
30. c.364,14,MSIF-20: WG to MS, 9 Apr. 1835.
31. c.371,21,MSP-2: Sir Robert Peel to MS, 30 Mar. 1835.
32. c.375,25,MSDIP-8: MS to Sir Robert Peel, 31 Mar. 1835 (copy).
33. c.375,25,MSDIP-8: King's Warrant, 1 Apr. 1835. Signed by Peel.
34. c.371,21,MSP-2: Sir Robert Peel to MS, 1 Apr. 1835.
35. c.361,11,MSFP-1: MS to WG, 13 July 1835.
36. c.361,11,MSFP-1: MS to WG, 15 July 1835.
37. Beinecke Rare Book and Manuscript Library, Yale University, Osborn Collection: Sir Robert Peel to Robert Southey, 4 Apr. 1835.
38. Jennings, *Croker Papers*, ii. 289.
39. Henry Bence Jones, *The Life and Letters of Faraday* (London, 1870), ii. 56–57.
40. *ibid.*, ii. 57.
41. 'Shewing How the Tories and the Whigs Extend Their Patronage to Science and Literature', *Frazier's Magazine*, xii (1835), 703–709.
42. Bence Jones, *Faraday*, ii. 58–59.
43. *ibid.*, ii. 59–60.
44. *ibid.*, ii. 63.
45. c.371,21,MSI-1: Robert Harry Inglis to MS, 10 Apr. 1835.
46. *Athenaeum*, 389 (1835), 280.
47. c.375,25,MSDIP-8: Robert E. Ferguson to WS, 5 Aug. 1836.
48. BL Addit. MSS. 37190 (Babbage Papers), f.204: MS to Charles Babbage, 6 June 1837.
49. c.375,25,MSDIP-8: Robert E. Ferguson to MS, 7 May 1837.
50. c.375,25,MSDIP-8: G. E. Anson to Robert E. Ferguson, 27 May 1837.
51. c.375,25,MSDIP-8: King's Warrant, 1837. Signed by Melbourne.
52. PR, p. 178.
53. *Athenaeum*, 502 (1837), 426.
54. Charles Babbage, *The Great Exhibition* (London, 1851), p. 189.
55. Hansard, 3rd. series, 1837, xxix, 1316.
56. *ibid.*, 1317.
57. c.362,12,MSIF-8: MS to Martha S, 22 Dec. 1837.
58. *The Times*, 20 Dec. 1837.

Chapter 9 — 'The Comet', An Experiment, and a Third Edition
pages 163–177

1. c.364,14,MSFP-20: WG to MS, 9 Apr. 1835.
2. c.358,8,MSFP-23: W. Rutherford, jun., to WG, 18 Apr. 1835.
3. c.358,8,MSFP-22: Mary Fairfax of Gillings Castle, near York, to MS, 20 Apr. 1835.
4. Moll, *Alleged Decline of Science*, p. 33.
5. c.371,21,MSM-4: Gerard Moll to WS, 14 Apr. 1835.
6. c.361,11,MSIF-1: MS to WG, 16 Mar. 1835.
7. Private communication, I. M. McCabe, Librarian, Royal Institution. Somerville's name appears on the List of 6 April and 18 April 1835.
8. c.369,19,MSA-1: D. F. J. Arago to MS, 8 Apr. 1835.

9. c.361,11,MSIF-1: MS to WG, 12 Apr. 1835.
10. Mary Somerville, *De la connexion des sciences physiques ou exposé et rapide de tous les principaux phènoménes astronomiques, physiques, chimiques, géologiques et météorologiques; accompagné des découvertes modernes, tant français qu'étrangers.* Traduit de l'Anglais, sous l'auspices de M. Arago, par Mme T. Meulien (Paris, 1839).
11. c.371,21,MSM-3: Marginal note in MS's aged hand on letter from Mme T. Meulien to MS, 26 Apr. 1835.
12. c.371,21,MSM-3: Mme T. Meulien to MS, 26 Apr. 1835.
13. c.375,25,MSDIP-3: S. S. Wayte, Secretary of the Bristol Institution, to MS, 5 Mar. 1835.
14. c.375,25,MSDIP-3: W. D. Coneybeare to MS, n.d. [March 1835].
15. *West of England Journal of Science and Literature,* iv (1835–36), 247–250.
16. *ibid.,* 249.
17. c.367,17,MSBY-1: Lady Noel Byron to MS, n.d. ('Monday').
18. c.370,20,MSC-3: Thomas Chalmers to MS, 16 June 1835.
19. c.369,19,MSB-12: H. I. Bowditch to MS, 20 Feb. 1835; 24 Feb. 1835; 16 Apr. 1835.
20. c.369,19,MSB-12: Nathaniel Bowditch to MS, 21 May 1835.
21. Ticknor, *Life,* i. 411, 412, 448–449, 479.
22. c.365,15,MSIF-34: Mary CS to WG, n.d. ('Thursday') [7 Mar. 1833].
23. PR, pp. 224–225.
24. [Dionysius Lardner], 'The Approaching Comet', *Edinburgh Review,* lxi (1835), 82–128.
25. c.361,11,MSIF-1: MS to WG, 13 July 1835.
26. c.373,23,MSBUS-3: John Murray to MS, 29 July n.y. [1835].
27. c.361,11,MSIF-1: MS to WG, 6 Aug. 1835; 8 Aug. 1835.
28. c.362,12,MSIF-8: MS to Martha S, 4 Aug. 1835.
29. c.362,12,MSIF-10: MS to Mary CS, 6 Sept. 1935.
30. c.372.22,MSS-6: Annarella Smyth (Mrs. W. H.) to MS, 4 Sept. 1835.
31. c.361,11,MSIF-1: MS to WG, 16 Sept. 1835.
32. *ibid.*
33. *Lit. Gaz.,* 970 (1835), 534.
34. c.362,12,MSIF-8: MS to Martha S, 27 Sept. 1835.
35. Dreyer, *History of the R.A.S.,* p. 64.
36. Mary Somerville, 'The Comet', *Quarterly Review,* lv (1835), 195–233.
37. *ibid.,* 218.
38. c.361,11,MSIF-1: MS to WG, 16 Sept. 1835.
39. Murray Archives: WS to John Murray, jun., n.d. ('Monday') [14 Sept. 1835].
40. c.362,12,MSIF-8: MS to Martha S, 27 Sept. 1835.
41. c.362,12,MSIF-8: MS to Martha S, 1 Oct. 1835.
42. c.362,12,MSIF-8: MS to Martha S, 23 Nov. 1835.
43. c.369,19,MSB-1: Charles Babbage to MS, 10 Oct. 1835.
44. c.362,12,MSIF-8: MS to Martha S, 11 Oct. 1835.
45. c.362,12,MSIF-8: MS to Martha S, 20 Oct. 1835.
46. c.372,22,MSS-6: W. H. Smyth to MS, 30 Sept. 1835; 3 Oct. 1835.
47. c.372,22,MSS-6: W. H. Smyth to MS, 5 Dec. 1835.
48. Murray Archives: WS to 'My dear Johnny', n.d. [Nov. 1835].
49. c.353,3,MSSW-11: Notes and working papers on Halley's comet.
50. c.369,19,MSB-5: Francis Beaufort to MS, 19 Nov. 1835.
51. W. S. Stratford, *On the Elements of the Orbit of Halley's Comet at its Appearance in the Years 1835–36* (London, 1836).
52. c.372,22,MSS-9: W. S. Stratford to MS, 6 Dec. 1835.
53. c.360,10,MSFP-54: Sir A. S. Greig to MS, 15 Oct. 1835.
54. c.362,12,MSIF-8: MS to Martha S, 28 Nov. 1835.
55. c.362,12,MSIF-8: MS to Martha S, 3 Dec. 1835.
56. John Cawood, 'The Magnetic Crusade: Science and Politics in Early Victorian Britain', *Isis,* 70 (1979), 493–518.
57. c.360,10,MSFP-54: P. Y. Fuss to Sir A. S. Greig, 12 Oct. 1835.

58. c.362,12,MSIF-8: MS to Martha S, 3 Dec. 1835.

59. Von Littow's *Ueber den Halleyschen Cometen* (Wien, 1835) and von Encke's *Ueber den Halleyschen Cometen* (Berliner Jahrbuch, 1835).

60. c.371,21,MSP-3: Joseph B. Pentland to WS, fragment, postmarked 22 Feb. 1835.

61. Francis Baily, *An Account of the Revd. John Flamsteed, the First Astronomer Royal* (London, 1835).

62. [Sir John Barrow], 'Account of the Rev. John Flamsteed', *Quarterly Review*, lv (1835), 568–572.

63. c.361,11,MSIF-1: MS to WG, 16 Sept. 1835.

64. c.371,21,MSP-2: Copy in MS's hand of letter from Sir Robert Peel, dated 30 Mar. 1835, with her note written much later.

65. c.361,11,MSIF-1: MS to WG, 8 July 1835.

66. c.362,12,MSIF-10: MS to Mary CS, 9 Aug. 1835.

67. c.361,11,MSIF-1: MS to WG, 30 Aug. 1835.

68. c.361,11,MSIF-1: MS to WG, 6 Aug. 1835.

69. c.361,11,MSIF-1: MS to WG, 8 July 1835.

70. c.361,11,MSIF-1: MS to WG, 13 July 1835.

71. c.361,11,MSIF-1: MS to WG, 15 July 1835.

72. c.361,11,MSIF-1: MS to WG, 13 July 1835.

73. c.361,11,MSIF-1: MS to WG, 15 July 1835.

74. c.362,12,MSIF-8: MS to Martha S, 25 July 1835.

75. c.362,12,MSIF-8: MS to Martha S, 12 Aug. 1835.

76. *The Examiner*, 1437 (1835), 534.

77. Lydia Tompkins, *Thoughts on the Ladies of the Aristocracy* (London, 1835).

78. c.362,12,MSIF-10: MS to Mary CS, 6 Sept. 1835.

79. Murray Archives: WS to John Murray, jun., 24 Aug. n.y. [1835].

80. c.361,11,MSIF-1: MS to WG, 17 Aug. n.y. [1835].

81. c.361,11,MSIF-1: MS to WG, 30 Aug. 1835.

82. c.361,11,MSIF-1: MS to WG, 16 Sept. 1835.

83. c.362,12,MSIF-8: MS to Martha S, 1 Oct. 1835.

84. c.370,20,MSF-1: Michael Faraday to MS, 12 Oct. 1835.

85. OCPS (1835), p. 245.

86. Bence Jones, *Faraday*, ii. 69–70.

87. c.352,2,MSSW-5: Drafts of letters describing 1835 experiments.

88. 'Experiments on transmission of the chemical rays of the solar spectrum across different areas. Excerpts from a letter of Mrs. Sommerville [*sic*] to Mr. Arago', *Comptes rendus*, iii (1836), 473–476.

89. c.361,11,MSIF-1: MS to WG, 15 July 1835; 12 Aug. 1835; 30 Aug. 1835.

90. c.361,11,MSIF-1: MS to WG, 12 Aug. 1835.

91. c.362,12,MSIF-8: MS to Martha S, 16 Aug. 1835.

92. c.362,12,MSIF-8: MS to Martha S, 1 Oct. 1835.

93. c.362,12,MSIF-7: MS to her daughters, 19 Dec. 1835.

94. c.358,8,MSFP-17: Richard Napier to MS, 20 Sept. 1835.

95. c.370,20,MSH-2: Henry Hallam to MS, 12 Mar. n.y. [1835].

96. J. B. Biot, *Recherches sur plusieurs points de l'astronomie égyptienne, appliquées aux monuments astronomiques trouvés en Égypte* (Paris, 1823).

97. Christ Church Library (Oxford), Hallam Papers: MS to Henry Hallam, 16 Mar. n.y. [1835].

98. c.370,20,MSF-2: W. H. Fitton to WS, 21 June n.y. [1835].

99. c.362,12,MSIF-10: MS to Mary CS, 6 Sept. 1835.

100. c.369,19,MSB-1: Charles Babbage to MS, 8 Oct. 1835.

101. c.369,19,MSB-1: Charles Babbage to MS, 20 Nov. n.y. [1835].

102. c.367,17,MSBY-3: Ada Augusta King to MS, 13 Dec. 1835.

103. c.362,12,MSIF-10: MS to Mary CS, 19 Dec. 1835.

104. c.361,11,MSIF-1: MS to WG, 24 Dec. 1835.

105. c.371,21,MSP-2: Sir Robert Peel to MS, 30 Mar. 1836.
106. c.361,11,MSIF-1: MS to WG, 6 Apr. 1835.
107. c.369,19,MSA-1: John Allen to MS, 4 May n.y. [1836].
108. Mary Somerville, *On the Connexion of the Physical Sciences* (3rd edn., London, 1836), pp. 22, 232.
109. c.373,23,MSBUS-1: Murray Accounts, Jan. 1836–6 July 1836.
110. c.361,11,MSIF-1: MS to WG, 1 Mar. 1836.
111. *Athenaeum*, 436 (1836), 181.
112. *Lit. Gaz.*, 997 (1836), 144.
113. *ibid.*, 1003 (1836), 286.
114. *Athenaeum*, 444 (1836), 318.
115. *Lit. Gaz.*, 1092 (1836), 751.

Chapter 10 — The Last London Years
pages 179–187

1. c.362,12,MSIF-8: MS to Martha S, 3 Jan. n.y. [1835].
2. c.361,11,MSIF-1: MS to WG, 1 Mar. 1836.
3. c.361,11,MSIF-1: MS to WG, 6 Apr. 1836.
4. c.351,1,MSSW-7: WS notes.
5. Babbage, *Reflections on the Decline of Science*, p. 107.
6. Charles Babbage, *Passages from the Life of a Philosopher* (London, 1864) (Dover edition, 1961), pp. 73–74.
7. c.369,19,MSB-1: Charles Babbage to WS, 16 Apr. 1836.
8. c.369,19,MSB-1: Charles Babbage to WS, 1 May 1836.
9. Babbage, *Ninth Bridgewater Treatise*.
10. c.365,15,MSIF-34: Mary CS to WG, 28 Mar. 1837.
11. BL Addit. MSS. 37190, f.204: MS to Charles Babbage, 6 June 1837.
12. c.362,12,MSIF-8: MS to Martha S, 30 Dec. 1836.
13. BL Addit. MSS. 37190, f. 385: MS to Charles Babbage, 11 July 1836.
14. c.369,19,MSB-1: Charles Babbage to MS, 12 July 1836.
15. *Lit. Gaz.*, 1023 (1836), 546.
16. c.361,11,MSIF-1: MS to WG, 6 Sept. 1836.
17. *ibid.*
18. c.365,15,MSIF-32: Martha S to WG, n.d. Oct. 1836.
19. c.370,20,MSH-6: Henry Holland to MS, n.d. ('Thursday') [18 Aug. 1836].
20. c.371,21,MSP-4: G. A. Plana to MS, 30 Oct. 1836.
21. c.371,21,MSP-4: J. A. F. Plateau to MS, 9 Nov. 1836.
22. c.372,22,MSW-3: J. G. Wilkinson to MS, n.d. [27? June 1836].
23. c.370,20,MSE-2: Montstuart Elphinstone to WS, 13 July 1836.
24. D. E. Borthwick, 'Outfitting the United States Exploring Expedition: Lieutenant Charles Wilkes' European Assignment, August-November 1836', *Proceedings of the American Philosophical Society*, 109 (1965), 159–172.
25. c.372,22,MSW-3: Charles Wilkes to MS, 18 Aug. 1838.
26. E. W. Farrar, *Recollections of Seventy Years* (Boston, MA, 1866), pp. 185–190.
27. E. W. Farrar, *The Story of the Life of Lafayette, as Told by a Father to his Children* (Boston, MA, 1831).
28. c.362,12,MSIF-8: MS to Martha S, 30 Dec. 1836.
29. *ibid.*
30. c.363,13,MSIF-18: WS to Martha S, 3 Jan. 1837.
31. c.362,12,MSIF-8: MS to Martha S, 30 Dec. 1836.
32. b.205,29,MSIF$_2$-61: MS to Jane Williamson Ramsay, 1 Jan. 1837.
33. *ibid.*
34. c.362,12,MSIF-8: MS to Martha S, 3 Jan. 1837.
35. c.363,13,MSIF-19: WS to WG, 22 Oct. 1832.

36. c.363,13,MSIF-18: WS to Martha S, 3 Jan. 1837.
37. *ibid.*
38. Joseph Henry, 'On the application of the principle of the galvanic multiplier to electromagnetic apparatus, and also to the development of great magnetic power in soft iron, with small galvanic elements', *American Journal of Science and the Arts,* xix (1831), 400.
39. OCPS (1834), p. 329.
40. OCPS (1835), p. 347; (1836), p. 332; (1837), p. 350; (1840), p. 350.
41. N. Reingold (ed.), *The Papers of Joseph Henry* (Washington, D.C., 1975), ii, 186.
42. N. Reingold (ed.), *The Papers of Joseph Henry* (Washington, D.C., 1979), iii, 257.
43. *ibid.,* iii. 292.
44. *ibid.,* iii. 292–293.
45. *ibid.,* iii. 360.
46. *ibid.*
47. Private communication, John Robert Murray to ECP, 17 Oct. 1979.
48. *Lit. Gaz.,* 1087 (1837), 741.
49. *ibid.,* 1092 (1837), 822.
50. c.373,23,MSBUS-2: Memoranda of numbers of books printed.
51. Private communication, John Robert Murray to ECP, 17 Oct. 1979.
52. Mary Somerville, *On the Connexion of the Physical Sciences* (4th edn., London, 1837).
53. *ibid.,* Section XXXV, 'Lexel's . . . Halley's', pp. 379–383.
54. *ibid.,* p. 120.
55. Williams, *Faraday Correspondence,* i. 306–307.
56. Richard Taylor (ed.), *Scientific Memoirs, Selected from the Transactions of Foreign Academies of Science and Learned Societies and from Foreign Journals* (London, 1837).
57. Bence Jones, *Faraday,* ii. 86.
58. c.370,20,MSF-1: Michael Faraday to WS, 10 Apr. 1837.
59. c.357,17,MSBY-2: Ada Augusta King to MS, 5 Feb. n.y. [1837].
60. Somerville College (Oxford), Somerville Memorabilia: MS to Prince Koslofski, 5 May 1837.
61. OCPS (1837), pp. 244–245.
62. *ibid.,* pp. 248–249.
63. *ibid.,* p. 269.
64. *ibid.,* p. 363.
65. c.370,20,MSF-2: R. W. Fox to MS, 30 Apr. 1836.
66. c.361,11,MSIF-2: MS to WG, 4 Mar. 1837.
67. c.363,13,MSIF-18: WS to 'My beloved daughters', 17 Sept. 1837.
68. R. L. Brett (ed.), *Barclay Fox's Journal* (London, 1979), p. 73.
69. Mrs. Stair Douglas, *The Life and Selections from the Correspondence of William Whewell, D.D.* (London, 1881), p. 437.
70. c.362,12,MSIF-10: MS to Mary CS, 19 Sept. 1837.
71. Howarth, *The British Association,* p. 174.
72. I. Todhunter, *William Whewell, D.D.* (London, 1876), ii. 201–202.
73. c.372,22,MSW-2: William Whewell to MS, 5 Jan. 1838.
74. c.370,20,MSF-20: R. W. Fox to MS, 10 Apr. 1836.
75. Trinity College (Cambridge) Library, Whewell Papers: MS to William Whewell, 6 Feb. 1838.
76. Trinity College (Cambridge) Library, Whewell Papers: MS to William Whewell, 20 Aug. 1838.
77. *Athenaeum,* 566 (1838), 637–638.
78. c.372,22,MSW-2: William Whewell to MS, 29 Aug. 1838.
79. Cawood, 'Magnetic Crusade', 493–518.
80. c.369,19,MSB-6: Mary Berry to MS, 6 Sept. 1838.
81. c.361,11,MSIF-2: MS to WG, 18 Sept. 1838.

Chapter 11 — Outside the Mainstream of Science
pages 189–195

1.   c.361,11,MSIF-2: MS to WG, 10 Dec. 1838.
2.   W. F. Moneypenny and G. E. Buckle, *The Life of Benjamin Disraeli, Earl of Beaconsfield* (new edn., London, 1929), i. 419.
3.   Ticknor, *Life*, ii. 54.
4.   c.355,5,MSAU-2: pp. 152–153.
5.   *Athenaeum*, 555 (1838), 423–426.
6.   *ibid.*, 558 (1838), 475.
7.   RS Herschel Papers, 16.404: J. F. W. Herschel to WS, 17 July 1835; 16.405: WS to J. F. W. Herschel, 28 July 1837.
8.   D. S. Evans, *Herschel at the Cape* (Austin, TX, 1969), pp. 176, 191, 237.
9.   c.369,19,MSB-12: H. I. Bowditch to MS, 5 May 1838.
10. PR, pp. 223–224.
11. c.361,11,MSIF-2: MS to WG, 10 Dec. 1838.
12. c.363,13,MSIF-13: MS to Agnes G. Greig (Mrs. Woronzow), 22 Dec. 1838.
13. c.361,11,MSIF-2: MS to WG, 23 Apr. 1839.
14. c.361,11,MSIF-2: MS to WG, 14 June 1839.
15. *Lit. Gaz.*, 1182 (1839), 590.
16. c.361,11,MSIF-2: MS to WG, 14 June 1839.
17. c.371,21,MSL-3: Grand Duke Leopold of Tuscany to MS, 31 May 1839.
18. c.361,11,MSIF-2: MS to WG, 14 June 1839.
19. c.355,5,MSAU-2: p. 165.
20. c.363,13,MSIF-13: MS to Agnes G. Greig (Mrs. Woronzow), 24 Oct. 1839.
21. BE 7192: WS to G. B. Amici, 25 Nov. 1839.
22. BE 7194: G. B. Amici to WS, n.d. [Nov. or Dec. 1839].
23. c.361,11,MSIF-2: MS to WG, 7 Jan. 1840.
24. BE 7204: MS to G. B. Amici, n.d. [winter or spring, 1840]; 7206: MS to G. B. Amici, n.d. [winter or spring, 1840].
25. Requested were 'Captn Beechey's voyage to the Pacific [F. W. Beechey, *Narrative of a voyage to the Pacific and Beering [sic] Strait, to co-operate with the polar expedition; performed in H.M. Ship Blossom under the command of Captain F. W. Beechey in the years 1825, 1826, 1827, 1828* . . . (London, 1831)]; the first volume of Bishoffs [sic] works [possibly G. W. von Bischoff's *Lehrbuch der botanik* (34 vols., Stuttgart, 1834–40)]; *Encyclopedia Britannica* article 'Rivers'; Marsdens history of Sumatra [W. Marsden, *History of Sumatra* (London, 1783)]; Raffles account of Java [T. S. Raffles, *History of Java* (London, 1817)]; Major Rennel's [sic] account of the Ganges [probably James Rennell's *An Account of the Ganges and Borranepooter Rivers. Read at the Royal Society, Jan. 25, 1780* (London, 1781)]'.
26. NLS 2213 f.145; MS to Mrs. Leonard Horner, 10 July 1840.
27. Mary Somerville, *On the Connexion of the Physical Sciences* (5th edn., London, 1840).
28. c.373,23,MSBUS-2: List of numbers of books in each edition and some costs.
29. c.373,23,MSBUS-3: Sheet showing royalties ('2/3 of profits').
30. *ibid.*
31. *The Times*, 13 Apr. 1840.
32. *ibid.*, 15 Apr. 1840.
33. NLS 2213 f.145; MS to Mrs. Leonard Horner, 10 July 1840.
34. BE 7195: WS to G. B. Amici, 7 Aug. 1840.
35. BL Addit. MSS 37190, f.369: MS to Charles Babbage, 2 May 1840.
36. NLS 2213 f.145: MS to Mrs. Leonard Horner, 10 July 1840.
37. BL Addit. MSS. 37190, f.370: MS to Charles Babbage, 13 Sept. n.y. [1840].
38. BE 7197: WS to G. B. Amici, 14 Oct. 1840.

39. c.363,13,MSIF-13: MS to Agnes G. Greig (Mrs. Woronzow), 6 Nov. 1840.
40. Mary Somerville, *Abstracts Phil. Trans.*, v (1843–1850), 569–570.
41. Mary Somerville, *Physical Geography* (London, 1848).
42. Mary Somerville, *On Molecular and Microscopic Science* (London, 1869).
43. c.370,20,MSD-1: Charles Darwin to MS, 21 Jan. n.y. [after 1862].

BIBLIOGRAPHY

Manuscript Sources

Cambridge
    Trinity College Library
        Whewell Papers

Claydon House, Buckinghamshire
        Verney Family Papers

Edinburgh
    National Library of Scotland
        Leonard Horner Papers
        Minto Papers
        John Thomson Papers

London
    British Library
        Babbage Papers
    John Murray (Publishers) Ltd., 50 Albemarle Street
        Papers of John Murray I and John Murray II in Murray Archives
    Royal Institution
        House List of Proposed Visitors
        Managers' Minutes
    Royal Society
        Certificate Ledgers
        J. F. W. Herschel Papers
        J. W. Lubbock Papers

Modena, Italy
    Biblioteca Estense
        Correspondence of G. B. Amici

New Haven, Connecticut, U.S.A.
    Beinecke Rare Book and Manuscript Library, Yale University
        Benjamin Silliman Papers
        Osborn Collection: Robert Southey letter
        Uncatalogued letters of Mary Somerville

Oxford
    Bodleian Library
        The Somerville Collection
    Christ Church Library
        Hallam Papers

Magdalen College Library
    Daubeny Letter Book
Somerville College
    Somerville Memorabilia
    Unpublished letters owned by Mrs. C. Colvin
    Edgeworth Family Papers

Paris
    Bibliothèque Nationale
    Bouvard Papers

St. Andrews, Scotland
    University of St. Andrews Library
    James David Forbes Papers

Printed Sources

Primary

Abstracts of the Papers Printed in the Philosophical Transactions of the Royal Society of London from 1800 to 1830 inclusive. Vol. ii (London, 1833).
Abstracts of the Papers Communicated to the Royal Society of London from 1843 to 1850, inclusive. Vol. v (London, 1851).
W. Airy (ed.), *Autobiography of Sir George Biddle Airy, K.C.B., etc.* (Cambridge, 1896).
The American Journal of Science and the Arts [Silliman's Journal] (New Haven, Connecticut, U.S.A.)
Annalen der Physik und Chemie [Poggendorff's Annalen] (Berlin).
Annales de chimie et de physique (Paris).
Annals of the Royal Statistical Society 1834–1934 (London, 1934).
The Army List [The Monthly Army List] (London, 1798–1940).
The Athenaeum (London).
Charles Babbage, *The Exposition of 1851; or, Views of the Industry, the Science, and Government of England* (London, 1851).
Charles Babbage, *The Ninth Bridgewater Treatise: A Fragment* (London, 1837).
Charles Babbage, *Passages from the Life of a Philosopher* (London, 1864) (Dover edition, 1961).
Charles Babbage, *Reflections on the Decline of Science in England and Some of its Causes* (London, 1830).
V. Y. Bowditch, *Life and Correspondence of Henry Ingersoll Bowditch* (2 vols., Boston, MA, 1902).
R. L. Brett (ed.), *Barclay Fox's Journal* (London, 1979).
The British Critic, Quarterly Theological Review, and Ecclesiastical Record (London).
William Buckland, *The Creation of the World, Addressed to R. I. Murchison and Dedicated to the Geological Society* (London, 1840).
Bulletin des sciences, mathématiques, astronomiques, physiques et chimiques (Paris).
J. P. T. Bury (ed.), *Romilly's Cambridge Diary 1832–42* (Cambridge, 1967).
J. W. Clark and T. McK. Hughes, *The Life and Letters of the Reverend Adam Sedgwick, LL.D., etc.* (2 vols., Cambridge, 1890).
William Cockburn, *A Letter to Professor Buckland* (London, 1838).

William Cockburn, *A Remonstrance, Addressed to His Grace the Duke of Northumberland, upon the Dangers of Peripatetic Philosophy* (London, 1838).

C. Colvin (ed.), *Maria Edgeworth — Letters from England 1813–1844* (Oxford, 1971).

Comptes rendus, iii (Paris).

B. B. Cooper, *The Life of Sir Astley Paston Cooper, Bart., Interspersed with Sketches from his Note-books of Distinguished Contemporary Characters* (2 vols., London, 1843).

J. F. Cooper, *A Residence in France; with an Excursion up the Rhine and a Second Visit to Switzerland* (Paris, 1836).

E. C. Curwen (ed.), *The Journal of Gideon Mantell, Surgeon and Geologist, covering the years 1818–1852, edited with an introduction and notes by E. Cecil Curwen* (London, 1940).

W. Daniell, *Sketches Representing the Native Tribes, Animals and Scenery of Southern Africa, from Drawings Made by the Late Samuel Daniell, Engraved by William Daniell* (London, 1820).

[Davenport], *Electro-Magnetism. History of Davenport's Invention of the Application of Electro-Magnetism to Machinery; with remarks on the same from the American Journal of Science and Arts, by Prof. Silliman. Also, extracts from other public journals, and information on electricity, galvanism, electro-magnetism, etc. by Mrs. Somerville* (New York, 1837).

Mrs. Stair Douglas, *The Life and Selections from the Correspondence of William Whewell, D.D.* (London, 1881).

Edinburgh New Philosophical Journal (Edinburgh).

Edinburgh Review (Edinburgh).

Encyclopaedia Britannica, Vol. xvii (3rd edn., Edinburgh, 1797).

Encyclopaedia Metropolitana, Vol. iv, Mixed Sciences (London, 1830).

The Examiner (London, 1835).

Michael Faraday, *Experimental Researches in Electricity* (Everyman's Library edition, London, 1914).

E. W. R. Farrar, *Recollections of Seventy Years* (Boston, MA, 1866).

[W. H. Fitton], *A Statement of the Circumstances Connected with the Late Election for the Presidency of the Royal Society* (London, 1831).

J. D. Forbes, *Testimonials in Favour of James David Forbes, F.R.S.S L. & E., F. G. S., Member of the Royal Geographical Society, of the Society of Arts for Scotland & Hon. Member of the Yorkshire Philosophical Society, as a Candidate for the Chair of Natural Philosophy in the University of Edinburgh* (Edinburgh, 1832).

Frazier's Magazine (London).

A. T. Gage, *History of the Linnean Society* (London, 1938).

Archibald Geikie, *Life of Sir Roderick I. Murchison* (2 vols., London, 1875).

Gentleman's Magazine (London).

W. E. Gladstone, *The Gladstone Diaries*, ed. M. R. D. Foot (2 vols., Oxford, 1968).

Mrs. [Elizabeth O. Buckland] Gordon, *The Life and Correspondence of William Buckland, D.D., F.R.S., Some Time Dean of Westminster, Twice President of the Geological Society, and First President of the British Association* (London, 1894).

Mrs. [Margaret Brewster] Gordon, *Home Life of Sir David Brewster, by his Daughter* (Edinburgh, 1870).

A. B. Granville, *The Royal Society in the XIXth Century, Being a Statistical Summary of its Labours . . .* (London, 1836).

A. B. Granville, *Science Without a Head, or, The Royal Society Dissected. By One of the 686 F.R.S.—sss* (London, 1830).

R. P. Graves, *Life of Sir William Rowan Hamilton* (3 vols., Dublin, 1882, 1885, 1889).

Charles C. F. Greville, *Memoirs 1814–1860*, ed. L. Strachey and R. Fulford (8 vols., London, 1938).

M. C. Harding (ed.), *Correspondance de H. C. Oersted avec divers savants* (2 vols., Copenhagen, 1920).

Hansard, *Parliamentary Debates*, 3rd series (London).

J. F. W. Herschel, *A Preliminary Discourse on the Study of Natural Philosophy* (London, 1830).

J. F. W. Herschel, *A Treatise on Astronomy* (London, 1830).

Mrs. John Herschel, *Memoir and Correspondence of Caroline Herschel* (London, 1876).

L. J. Jennings (ed.), *The Correspondence and Diaries of the Late Rt. Hon. J. W. Croker* (2nd edn. revised, 3 vols., London, 1885).

George Jones, *Sir Francis Chantrey: Recollections of his Life, Practice & Opinions* (London, 1849).

Henry Bence Jones, *The Life and Letters of Faraday* (2 vols., London, 1870).

Journal des Savans (Paris).

Andrew Lang, *Life and Letters of John Gibson Lockhart* (2 vols., London, 1897).

[Dionysius Lardner], 'The Approaching Comet', *Edinburgh Review*, lxi (1835), 82–128.

List of Carthusians 1801 to 1879 (London, 1879).

List of the Linnean Society of London (London, 1817; 1829; 1830).

J. G. Lockhart, *Narrative of the Life of Sir Walter Scott, Bart., Begun by Himself & Continued by J. G. Lockhart* (2 vols., Edinburgh and London, 1848).

The London Literary Gazette; and Journal of Belles Lettres, Arts, Sciences, &c. [*Literary Gazette*] (London).

London Medical Gazette, iii (1829).

Katharine M. Lyell, *Memoir of Leonard Horner, F.R.S., F.G.S., Consisting of Letters to His Family and from Some of His Friends* (2 vols., London, 1890).

Mrs. [Katharine M.] Lyell (ed.), *Life, Letters and Journals of Sir Charles Lyell, Bart.* (2 vols., London, 1881).

Harriet Martineau, *Biographical Sketches 1852–1868* (London, 1869).

Harriet Martineau, *Harriet Martineau's Autobiography with Memorials by Marcia Weston Chapman* (3 vols., London, 1877).

J. C. Maxwell, *The Scientific Papers of J. Clerk Maxwell*, ed. W. D. Niven (2 vols., Cambridge, 1890).

E. C. Mayne, *Life and Letters of Anna Isabella, Lady Noel Byron* (London, 1929).

Mechanics Magazine (London).

[Gerard Moll], *On the Alleged Decline of Science in England. By a Foreigner* (London, 1831).

Monthly Review (London).

W. F. Moneypenny and G. E. Buckle, *The Life of Benjamin Disraeli* (2 vols., London, 1929).

The Morning Post (London).

W. Munk, *The Roll of the Royal College of Physicians of London* (2nd edn., 3 vols., London, 1878).

New Monthly Magazine (London).

C. S. Parker, *Sir Robert Peel* (2 vols., London, 1899).

George Peacock, *Life of Thomas Young M.D., F.R.S., &c. and one of the Eight Foreign Associates of the National Institute of France* (London, 1855).

D. Peterkin and W. Johnstone, *Commissioned Officers in the Medical Services of the British Army, 1660–1960* (London, 1968).

Philosophical Transactions of the Royal Society of London (London).

John Playfair, *Outlines of Natural Philosophy, being the heads of lectures delivered at the University of Edinburgh* (3rd edn., Edinburgh, 1819).

Quarterly Review (London).

The Record of the Royal Society of London (London, 1901).

Registre de Proces-Verbaux et Rapports des Séances de l'Académie Royale des Sciences, x, Part i (1832).

Nathan Reingold (ed.), *The Papers of Joseph Henry*. Vol. 2 — November 1832–December 1835: The Princeton Years (Washington, D.C., 1975).

Nathan Reingold (ed.), *The Papers of Joseph Henry*. Vol. 3 — January 1836–December 1837: The Princeton Years (Washington, D.C., 1979).

John Robison, 'Physics', *Encyclopaedia Britannica* (3rd edn., Edinburgh, 1797).

Scots Magazine (Edinburgh).

J. C. Shairp, P. G. Tait, and A. Adams-Reilly, *Life and Letters of James David Forbes, F.R.S.* (London, 1873).

Samuel Smiles, *A Publisher and His Friends. Memoir and Correspondence of the Late John Murray, with an Account of the Origin and Progress of the House, 1768–1843* (2 vols., London, 1891).

Martha Somerville, *Personal Recollections from Early Life to Old Age of Mary Somerville, with Selections from her Correspondence* (London, 1873).

Mary Somerville, *De la connexion des sciences physiques ou exposé et rapide de tous les principaux phènoménes astronomiques, physiques, chimiques, géologiques, et météorologiques, accompagne des découvertes modernes, tant français qu'étrangers. Traduit de l'Anglais, sous l'auspices de M. Arago, par Mme T. Meulien* (Paris, 1839).

Mary Somerville, 'Experiments on transmission of the chemical rays of the solar spectrum across different areas. Excerpts from a letter of Mrs. Sommerville's [*sic*] to Mr. Arago', *Comptes rendus*, iii (1836), 473–476.

Mary Somerville, *The Mechanism of the Heavens* (London, 1831).

Mary Somerville, *On Molecular and Microscopic Science* (London, 1869).

Mary Somerville, *On the Connexion of the Physical Sciences* (London, 1834, 1835, 1836, 1837, 1840).

Mary Somerville, 'On the magnetizing power of the more refrangible solar rays'. Communicated by W. Somerville, M.D., F.R.S., Feb. 2, 1826', *Philosophical Transactions of the Royal Society of London*, cxvi (1826), 132.

Mary Somerville, *Physical Geography* (2 vols., London, 1848).

Mary Somerville, *Preliminary Dissertation to the 'Mechanism of the Heavens' by Mrs. Somerville* (London, 1831).

Mary Somerville, *A Preliminary Dissertation on the Mechanism of the Heavens* (Philadelphia, 1832).

Mary Somerville, 'Art. VII. — 1. Ueber den Halleyschen Cometen. Von Littrow. Wien, 1835. 2. Ueber den Halleyschen Cometen. Von Professor von Encke. Berliner Jahrbuch, 1835 &c. &c. &c.', *Quarterly Review*, lv (December, 1835), 195–223.

William Sotheby, *Lines Suggested by the Third Meeting of the British Association for the Advancement of Science, Held at Cambridge in June, 1833. By the Late*

William Sotheby, Esq. F.R.S., &c.&c. With a Short Memoir of his Life (London, 1834).

Sir James South, *Charges Against the President and Councils of the Royal Society* (2nd edn., London, 1830).

Sir Jame South, *On the Proceedings of the Royal Society, &c., the Necessity of a Reform of its Conduct, and a re-modelling of its Charter, &c.* (London, 1830).

Spectator (London).

Sydney Spokes, *Gideon Algernon Mantell, Surgeon and Geologist* (London, 1927).

Lydia Thompkins, *Thoughts on the Ladies of the Aristocracy* (London, 1835).

George Ticknor, *Life, Letters and Journals of George Ticknor* (2 vols. Boston, MA, 1876).

The Times (London).

A. C. Todd, *Beyond the Blaze: A Biography of Davies Gilbert* (Truro, 1967).

I. Todhunter, *William Whewell, D.D. Master of Trinity College, Cambridge. An Account of His Writings with selections from his Literary & Scientific Correspondence* (2 vols., London, 1876).

John Toplis, *A Treatise Upon Analytical Mechanics: Being the First Book of the Mecanique Celeste of P. S. Laplace* (Nottingham, 1814).

Transactions of the Royal Society of Edinburgh (Edinburgh).

J. A. Venn, *Alumni Catabrigensis. Part II. 1751–1900* (6 vols., Cambridge, 1940–1954).

West of England Journal of Science and Literature (Bristol, 1835–36).

L. P. Williams (ed.), *The Selected Correspondence of Michael Faraday* (2 vols., Cambridge, 1971).

Thomas Young, *Elementary Illustrations of the Celestial Mechanics of Laplace, with Some Additions Relating to the Motions of Waves and of Sound and to the Cohesion of Fluids* (London, 1821).

Thomas Young, *A Course of Lectures on Natural Philosophy and the Mechanical Arts* (2 vols., London, 1807).

Zeitschrift für Physik und mathematik (Bonn).

Secondary

Sir John Barrow, *Sketches of the Royal Society and the Royal Society Club* (London, 1849).

Sir John Barrow, *A Voyage to Cochin China* (London, 1806).

Morris Berman, *Social Change and Scientific Organization: The Royal Institution, 1790–1844* (Ithaca, NY, 1978).

D. E. Borthwick, 'Outfitting the United States Exploring Expedition: Lieutenant Charles Wilkes' European Assignment, August–November 1836', *Proceedings of the American Philosophical Society*, 109 (1965), 159–172.

S. F. Cannon, *Science in Culture* (New York, 1978).

D. S. L. Cardwell, *The Organization of Science in England* (London, 1972).

John Cawood, 'The Magnetic Crusade: Science and Politics in Early Victorian Britain', *Isis*, 70 (December 1979), 493–518.

Maurice Crosland, 'Development of the Professional Career in Science in France', in Maurice Crosland (ed.), *The Emergence of Science in Western Europe* (New York, 1976), pp. 139–159.

Maurice Crosland, *Gay-Lussac: Scientist and Bourgeois* (Cambridge, 1978).

Maurice Crosland, *The Society of Arcueil* (London, 1967).

Maurice Crosland and Crosbie Smith, 'The Transmission of Physics from France to Britain: 1800–1840', *Historical Studies in the Physical Sciences,* ix (Baltimore, 1978), pp. 1–61.

M. J. Cullen, *The Statistical Movement in Early Victorian Britain: The Foundation of Empirical Social Research* (New York, 1975).

C. G. T. Dean, *The Royal Hospital of Chelsea* (London, 1950).

Dictionary of National Biography (Compact edn., Oxford, 1975).

Dictionary of Scientific Biography (New York, 1969–1980).

Dictionary of South African Biography (Johannesburg, 1977).

C. R. Dodd, *A Manual of Dignities, Privileges and Precedence* (London, 1842).

J. E. Dreyer, H. H. Turner *et al., History of the Royal Astronomical Society 1820–1920* (London, 1923).

D. I. Duveen and H. S. Klickstein, *A Bibliography of the Works of A. L. Lavoisier* (London, 1954).

H. R. Fletcher, *The Story of the Horticultural Society 1804–1968* (London, 1969).

D. S. Evans, *Herschel at the Cape* (Austin, TX, 1969).

Alfred Friendly, *Beaufort of the Admiralty: The Life of Sir Francis Beaufort 1774–1857* (London, 1977).

Sir Archibald Garrod, 'Alexander John Gaspard Marcet', *Guy's Hospital Reports* (October, 1925).

Sir Archibald Geikie, *Annals of the Royal Society Club: The Record of a London Dining Club in the Eighteenth and Nineteenth Centuries* (London, 1917).

L. F. Gilbert, 'Election to the Presidency of the Royal Society in 1820', *Notes and Records of the Royal Society of London,* ii (1954–55).

Mollie Gillen, *Royal Duke: Augustus Frederick, Duke of Sussex (1773–1843)* (London, 1976).

Hudson Gurney, *Memoir of the Life of Thomas Young, M.D., F.R.S., &c. with a Catalogue of his Works and Essays* (London, 1831).

Sir Harold Hartley, *Humphry Davy* (London, 1966).

Henry Hasted, 'Reminiscences of Dr. Wollaston', *Proceedings of the Bury and West Suffolk Archeological Institute,* i (1849).

J. N. Hays, 'Science and Brougham's Society', *Annals of Science,* xx (Nov. 1965), 227–241.

O. J.R. Howarth, *The British Association for the Advancement of Science: A Retrospect 1831–1931* (2nd edn., London, 1931).

A. Hume, *The Learned Societies and Printing Clubs of the United Kingdom: Being an Account of their Respective Origin, History, Objects and Constitution: with full details respecting Membership Fees, their Published Works and Transactions, Notices of their Periods and Places of Meeting, &C. and a General Introduction and a Classified Index* (London, 1853).

A. Lacroix, 'Une famille de bons serviteurs de l'Académie des Sciences et du Jardin des Plantes', *Bulletin du Muséum,* 2^0 Serie, Tome X, No. 5 (1938).

Mrs. [Sarah W. Bowdich] Lee, *Memoir of Baron Cuvier* (London, 1833).

Sir Henry Lyons, *The Royal Society 1660–1940 — A History of its Administration under its Charters* (London, 1968).

L. Marchand, *The Athenaeum: A Mirror of Victorian Culture* (Chapel Hill, NC, 1941).

Jack Morrell and Arnold Thackray, *Gentlemen of Science: Early Years of the British Association for the Advancement of Science* (Oxford, 1981).

R. S. Morris, *Cholera 1832: The Social Response to an Epidemic* (New York, 1976).

Maboth Moseley, *Irascible Genius: A Life of Charles Babbage, Inventor* (London, 1964).

F. J. Mouat, *History of the Statistical Society of London* (London, 1885).

E. C. Patterson, *John Dalton and the Atomic Theory: The Biography of a Natural Philosopher* (New York, 1970).

Jasper Ridley, *Lord Palmerston* (London, 1970).

Robert Robson (ed.), *Ideas and Institutions of Victorian Britain* (London, 1967).

Crosbie Smith, '"Mechanical Philosophy" and the Emergence of Physics in Britain 1800–1850', *Annals of Science,* xxxiii (1976), 3–29.

E. F. Smith, *Old Chemistries* (New York, 1927).

H. R. Tedder, *The Athenaeum 1824–1934* [Reprinted from *The Times* of 16 Feb. 1924] (London, 1924).

A. Vucinich, *Science in Russian Culture: A History to 1860* (Stanford, CA, 1963).

F. C. Waugh, *Founding Members of the Athenaeum Club from its Foundation* (London, 1894).

C. R. Weld, *History of the Royal Society with Memoirs of the Presidents Compiled from Authentic Documents* (2 vols., London, 1848).

L. P. Williams, *Michael Faraday: A Biography* (New York, 1965).

L. P. Williams, 'The Royal Society and the Founding of the British Association for the Advancement of Science', *Notes and Records of the Royal Society,* xvi (1961), 221–233.

L. G. Wilson, *Charles Lyell — The Years to 1841: The Revolution in Geology* (New Haven, CT, 1972).

H. B. Woodward, *The History of the Geological Society of London* (London, 1907).

Philip Ziegler, *King William IV* (London, 1971).

INDEX

O. J. R. Howarth (2nd edn., London, 1931), 216, 224, 233
British Critic, Quarterly Theological Review and Ecclesiastical Record, 140, 217, 228
British Library, 197, 227
British Museum, 15, 99, 167
The British Review, 85
British science — see English science
British universities, ix, 49, 69, 89, 91–92, 93, 112–113, 132, 175
Brochant de Villiers, André Jean François Marie, 23, 25
Brockedon, William, 121
Broderip, William John, 43, 62, 99, 206–207, 212
Brodie, Benjamin Collins, 15, 52, 62, 205–207
Broglie, Albertine de Staël, Duchesse de Broglie (1797–1838), 97
Bromhead, Edward Ffrench (F.R.S., 1817), 207
Brongniart, Alexandre, 22, 116
Brooke, Henry James (1771–1857; F.R.S., 1819), 207
Brougham, Henry, Lord Brougham and Vaux, 5, 48–50, 70–71, 72–73, 75, 79–81, 83, 84, 88, 120, 130–131, 151, 152, 153, 157–158, 163, 205, 209–210, 216
Brougham, Miss (sister of H. Brougham), 5
Broughton, Samuel Daniel, 109
Brown, Robert, 15, 99, 207
Brunel, Isambard Kingdom, 42, 43, 203, 207
Brunel, Mark Isambard, 42, 207
Brussels, 10, 45, 59, 105
Buckland, Mary Morland (Mrs. William Buckland), 53, 115, 146
Buckland, William, 48, 52–53, 65, 68, 71, 77, 79, 92, 94, 112, 117, 126, 205, 207, 211, 228, 229
The Creation of the World, Addressed to R. I. Murchison and Dedicated to the Geological Society by William Buckland (London, 1840), 205, 208
The Life and Correspondence of William Buckland, D.D., F.R.S. by Mrs. Gordon (London, 1894), 94, 207, 211, 229
Buckle, G. E., 225, 230
Buller, Charles, 84, 161–162
Buller, Mr., 83
Bulletin des sciences, mathématiques, astronomiques, physiques et chimiques, 47–48, 204, 228
Bulletin du Muséum, 201, 233
'Bumps of Mathematics', 108
Burlington House (London), 60, 62

Burntisland (Fife), 1–2, 4, 5, 160, 170
Bury, J. P. T., 214, 217, 228
Butler, Richard, 203
Byron, Ada Augusta — see Lady Lovelace
Byron, George Gordon, Lord Byron (1788–1824), 29, 73, 142, 148, 149, 177
Byron, Anne Isabella Milbanke, Lady Noel Byron (1792–1860), 148–149, 219, 230
Life and Letters of Anne Isabella, Lady Noel Byron by E. C. Mayne (London, 1929), 219, 230

Cabinet Cyclopaedia, 119
Cabinet of Mineralogy (Muséum d'Histoire Naturelle, Paris), 21
Caesar's *Commentaries,* 2
Calais, 19
calculating engine, 55, 56, 88, 130, 180
calculating machine — see calculating engine
Callcott, Maria Dundas Graham, Lady Callcott (1785–1842), 74, 90
Callet's logarithms, 9
calls (visits), 12, 38, 41, 95, 103
calorific rays of the solar spectrum, 173, 184
Cambridge, 63, 68, 76, 88, 89, 91, 92, 93, 99, 109, 111, 115, 116, 117, 130, 138, 142, 150, 151, 152, 154, 164, 184
Cambridge by-election (1832), 108–109
Cambridge Philosophical Society, 70, 88, 91, 164
Cambridge University, 35, 44, 49, 51, 55–57, 63, 68, 89, 91, 109, 120, 126, 148, 150, 154
Cambridge University Observatory, 91, 154
Camden Hill Observatory, 168
camera lucida, 28
Cameron, Miss (niece of Mrs. Hudson Gurney), 147–148
Campbell, Thomas (1777–1844), 142
Camperdown, Battle of, 3, 153
Canada, 6, 7
Candolle, Augustin Pyramus de, 26, 29, 30, 120, 202
Cannon, S. F., 214, 232
Cape Colony, 6–7, 17, 130, 168, 174, 189
Cape of Good Hope — see Cape Colony
Cardwell, D. S. L., 232
Carey and Lea, Publishers (Philadelphia), 89
Carlists, 95, 104
Caroline, Queen (1768–1821), 50
catalytic action of platinum, 134
catatropic microscope, 28, 30–31
Catherine II, Empress of Russia (1729–1796), 3
Catholic emancipation (1829), 160